Glimpses of
Agricultural Extension

THE AUTHORS

Dr. R. K. Doharey is Professor & Head, Deptt of Ext. Edu., and Former Assoc Dean, College of Agriculture Campus, Kotwa, Azamgarh under ANDUA&T, Kumarganj, Ayodhya (UP). He passed B.Sc. Ag., (Hons.) from Kanpur University, Kanpur, UP in 1995, M.Sc. Ag. (Ext. Edu.) in 1997 from IAS, B.H.U., Varanasi and Ph.D. (Agril. Ext) from CSJMU, Kanpur in 2004. Qualified NET from ASRB, ICAR in 1997, 1999 and 2001. He is serving with wide exposure more than 23 years in different institutions/SAUs i.e. UPCAR, Lucknow from 1999-2001, CSAUA&T, Kanpur from 2001-2004 and ANDUA&T, Kumarganj, Ayodhya from 2005-to till now. He is as Professor of Ext. Edu. in more than 12 years. He holed many responsible positions of the University/Departments/ Institutions i.e Science Writer, TA/SMS (Ag. Ext.), TO/PC, KVKs, Assoc. Prof., Prof. (Ext.Edu.), Assoc. Dean, A.O., APIO, PIO, SVC, Head, Deptt of Ext. Edu., Head, Deptt. of Fisheries Ext., Economics and Statistics, CFS. He is involved with Teaching, Research, Extension. He guided 71 M.Sc. (Ag.) and 21 Ph.D Students. He had wide International Exposure visited 18 countries like United State of Emirates UAE (Dubai, Sharjah, Abu Dhabi), Doha (Qatar), United Kingdom (UK), Holland, Poland, Germany, Switzerland, Italy, France, Netherlands, Rome, Russia, Singapore, Malaysia, Thailand and Nepal etc. He has Published 115 Research Papers in National and International Journals, 146 research abstracts, 31 awards received from national and international societies/departments, delivered 56 TV talk, 33 Radio talk, attained 69 seminars/symposia, 17 Life members of the different Societies, paper setter of 29 universities/commissions, published 12 Books,12 Book Chapters, 43 Popular Article, 06-Folders, organized 21 Kisan Mela, compiled 09 Reports and handled 02 project etc.

Mr. Vishal Yadav has completed his B.Sc. (Hons.) Agriculture from ANDUA & T, Kumarganj, Ayodhya. He is currently in M. Sc. Scholar at Deptt. of Agril. Ext. Edu., ANDUA&T, Kumarganj, Ayodhya. He has cleared ICAR's JRF. He is a founder of youtube channel in Agri. Social Sciences by Vishal, and he also operate telegram channel ICAR JRF Social Science, ICAR SRF/NET Extension Education.

Dr. Satyapriya Currently employed at Division of Agril. Ext. at ICAR-IARI, New Delhi, as PS (Agril. Ext.). Having experience of 18 years of research, educating postgraduate students, conducting training programmes, and Agril. Ext. work in the field. He is gold medalist during PG studies at BHU. Earned Doctorate degree from BHU, Varanasi.

Dr. N.R. Meena is an Asstt. Prof., Department of Ext. Edu., College of Agriculture, ANDUA&T, Kumarganj, Ayodhya, Dr. Meena has 5 years' experience in teaching, Research and Extension at the University level. He authored 1book and published more than 20 research papers in National & International Journals. He has awarded Young Scientists award from ISEE, IARI, New Delhi and 01 Excellence in Teaching award from Astha Foundation, Meerut, UP.

Dr. Manoj Kumar had obtained his M.Sc. (Ext. Edu.) from ANDUA&T, Kumarganj, Ayodhya and Ph.D. in Agril. Ext. and Communication from SVPUA&T, Meerut . Currently he is working as SMS (Agril. Ext.) in KVK, Belipar, Gorakhpur under ANDUAT, Kumarganj, Ayodhya. (UP).

Dr. Ashwani Kumar Verma is working as Asstt. Prof., Division of Agril. Ext., RBS College, Agra. He has cleared ICAR-and UGC-NET. He has authored of 02 books 07 research papers in various reputed journals and 14 popular articles published in ICAR/SAUs magazines.

Glimpses of
Agricultural Extension

For UG/PG/ICAR JRF SRF/NET BHU, UPHESC and Other Competitive Exam

by

R.K. Doharey

Vishal Yadav

Satyapriya

N.R. Meena

Manoj Kumar

Ashwani Kumar Verma

2023

Astral Cracker

A Division of

Astral International Pvt. Ltd.

New Delhi – 110 002

ISBN: 978-93-5461-670-9 (PB)

Publisher's Note:

Every possible effort has been made to ensure that the information contained in this book is accurate at the time of going to press, and the publisher and author cannot accept responsibility for any errors or omissions, however caused. No responsibility for loss or damage occasioned to any person acting, or refraining from action, as a result of the material in this publication can be accepted by the editor, the publisher or the author. The Publisher is not associated with any product or vendor mentioned in the book. The contents of this work are intended to further general scientific research, understanding and discussion only. Readers should consult with a specialist where appropriate.

Every effort has been made to trace the owners of copyright material used in this book, if any. The author and the publisher will be grateful for any omission brought to their notice for acknowledgement in the future editions of the book.

Published by : **Astral Cracker**
A Division of
Astral International Pvt. Ltd.
– ISO 9001:2015 Certified Company
4736/23, Ansari Road, Darya Ganj
New Delhi-110 002
Ph. 011-43549197, 8130496929
E-mail: info@astralint.com
Website: www.astralint.com

Digitally Printed at : **Replika Press Pvt. Ltd.**

Dedicated to our

BELOVED PARENTS

भारतीय कृषि अनुसंधान परिषद
कृषि अनुसंधान भवन–1, पूसा, नई दिल्ली 110 012
INDIAN COUNCIL OF AGRICULTURAL RESEARCH
Krishi Anusandhan Bhawan, Pusa, New Delhi – 110012
Ph.:91-11-25843277 (O), Fax : 91-11-25842968
E-mail: aksicar@gmail.com

डा. अशोक कुमार सिंह
उप महानिदेशक (कृषि प्रसार)
Dr. A.K. Singh
Deputy Director General (Agricultural Extension)

Foreword

Agricultural extension helps to discover needs, problems and build programmes to achieve desired goals. In the changing agricultural scenario, the role of extension has broadened looking into challenges that are currently being faced. Research outcomes may not be generated every day, but extension system ought to be vibrant to meet out requirements of diverse changes happening around.

The book entitled "Glimpses of Agricultural Extension" has been written with the purpose of fulfilling the needs of the students, research scholars, teachers, planners, administrators, extension workers *etc.* who are engaged in teaching and doing research on agricultural extension. Recently the Indian Council of Agricultural Research (ICAR) has revised the syllabus for the graduate students as per report of 5th Dean Committee. In the agricultural education courses, the Extension Education is an important course for under-graduation and post-graduation degree programme. The present book consists of ten sections which contains specific details about various concepts, methods, programmes and issues.

I am happy to find that "Glimpses of Agricultural Extension" has been prepared by teachers of Department Extension Education. I believe that manual will be helpful to the students. I congratulate Dr. R. K. Doharey, Vishal Yadav, Dr. Satyapriya, Dr N. R. Meena, Dr. Manoj Kumar and Dr. Ashwani Verma for preparing this Book.

(A.K. Singh)

Dated :28.06.2022

Mahatma Gandhi University of Horticulture & Forestry

Sankra-Patan, Distt – Durg - 491111 Chhattisgarh

E-mail – mguvvdurg@gmail.com, mguvv-durg@cg.gov.in

Ref No. - MGUVV/Estt./2022/240 **Dated: 23.06.22**

Foreword

In the changing agricultural scenario, the role of extension education has broadened looking into challenges that are currently being faced. Research outcomes may not be generated every day, but extension system ought to be vibrant to meet out requirements of diverse changes happening around. Agricultural extension helps farmers to discover their needs, problems and build educational programmes based on their needs.

A This Book Glimpses of Agricultural Extension Education has been written for fulfilling the needs of the students, research scholars, teachers, planners, administrators, extension workers etc, who are engaged in teaching, research and agricultural extension education. Indian Council of Agricultural Research (ICAR) has revised the syllabus for the graduate as per 5th Dean Committee recommendations recently. This syllabus is made common for all under graduate programmes in State Agricultural Universities in India. In agricultural courses curriculum, the Extension Education is important and essential courses in under-graduation and post-graduation degree programmes. In this book the information are divided in ten sections namely; Extension Concept & Scope, Extension Schemes, Communication, Adoption & Diffusion, Information Technology, Rural Sociology & Educational Psychology, Programme Planning, Management & Training, Entrepreneurship and Gender Sensitization & Empowerment.

I am pleased to know that a "Glimpses of Agricultural Extension Education" has been prepared by faculty of Department of Extension Education, Acharya Narendra Deva Univdrsity of Agriculture & Technology, Kumarganj, Ayodhya and I congratulate Dr. R. K. Doharey, Mr. Vishal Yadav, Dr. Satyapriya, Dr N. R. Meena, Dr. Manoj Kumar and Dr. Ashwani Verma for preparing this very useful manual for transforming extension education in india.

(R. S. Kureel)
Vice Chancellor

Formerly:

- Director: Institute of Horticulture Technology, Greater NOIDA.
- Member: National Education Policy-2020, Ministry of Education, Govt. of India.
- Vice Chancellor: Birsa Agriculture University , Ranchi Jharkhand
- Vice Chancellor: Dr. B.R. Ambedkar University of Social Science, MHOW Indore, M.P.
- Vice Chancellor: Acharya N.D. University of Agriculture & Technology Ayodhya U.P.
- Director: Ministry of Agriculture & Farmers welfares, Govt. of India

Preface

The book covers all the chapters of Agricultural Extension Education. While compiling this book various text books, reference books and syllabus of fifth Dean's committee of ICAR has been referred from which matters have been heavily drawn. We acknowledge our humble indebtedness to all those authors from whose work materials have been drawn. We express our sincere, heartfelt, gratitude to our beloved parents for their everlasting inspiration, abundant love, guidance and help.

Any suggestions for improvement or comment on the book are welcome; one can contact us at vishalyadav9220@gmail.com and rkdoharey@yahoo.co.in.

R.K. Doharey

Vishal Yadav

Satyapriya

N.R. Meena

Manoj Kumar

Ashwani Kumar Verma

Contents

Chapter 1

Important Books in Extension Education and Communication

- Challenging the professionals: **R. Chambers**
- Extension Education: **Advi Reddi**
- Agricultural Extension: **Van Den Ban**
- Extension Science: **N. Rolling**
- Extension communication and management: **G.L. Ray**
- Introduction to Extension education: **S.V. Supe**
- Extension education in community development: **J. Paul Leagans**
- Audio: Visual methods in training: **Edgar Dale**
- An introduction to Agricultural Extension: **A.T. Mosher**
- Foundation of behavioural research: **F.N. Kerlinger**
- A theory of human motivation: **Abraham H. Maslow**
- Extension teaching methods: **Mc Willson and G. Gallup**
- Organizational communication: **E. M. Rogers and Rekha Rogers**
- Communication of Innovation: **E.M. Rogers and E.F. Shoemakers**
- Diffusion of innovation: **E.M. Rogers**
- Knowledge and use of extension methods-a guide to community Development: **D. Ensminger**
- Study of +ve philosophy: **Augustecomte**

- ☆ Principles of philosophy: **Herbert Spencer**
- ☆ The human side of enterprise: **D. McGregor**
- ☆ Methods of social survey and research: **S.R. Bajpai**
- ☆ Pedagogy of the oppressed: **Paulo Freire**
- ☆ Traditional Media in India**: S.C. Parmer**
- ☆ Guide on alternative extension approaches**: G.H. Auxin**
- ☆ Cooperative extension work: **Kelsey and Herne**
- ☆ Cooperative extension service: **Sanders**
- ☆ Education and communication for development: **O.P. Dhama and O.P. Bhatnagar**
- ☆ Farmers first: A practical paradigm in rural development: **R. Chambers**
- ☆ Communication process in Rural development**: J.P. Leagans**
- ☆ The process of communication**: David K. Berlo**
- ☆ Introductory rural development: **J.B. Chitamber**
- ☆ Modernizing world agriculture**: S.K. Wagmare**
- ☆ Scientific Social Survey and Research**: P.V. Young**
- ☆ Attacking rural poverty: **Coomb and Ahmed**

IMPORTANT TERMS IN EXTENSION EDUCATION

- ☆ Social norms: **Layman**
- ☆ Folkways: **Sumner**
- ☆ Androgogy: **Alexander knapp**
- ☆ Father of Adult education in USA: **Lindmen**
- ☆ Motto of 4H club: **Carri Harrison**
- ☆ Folkways : **Sumner**
- ☆ POSDCORB: **Gullick**
- ☆ PRA : **Henry ford**
- ☆ RRA : **Collinson**
- ☆ Concept of self : **Cooley**
- ☆ Role taking theory: **G H Mead**
- ☆ Entrepreneur word: **Richard Cantinlon**
- ☆ Credibility: **Hovland**
- ☆ Sociometric technique: **J L Moreno**
- ☆ Polygynandry: **Majumadar**

- ☆ Urban fringe : **Srinivas**
- ☆ Social physiology : **Saint Simon**
- ☆ Term Stem family : **F Leplay**
- ☆ Concept of ego and Alter : **Talcott Parson**
- ☆ Concept of IQ : **Willium Stung**
- ☆ Conflict theory : **Karl Marx**
- ☆ Communism theory : **Karl marx**
- ☆ Disequilibrium theory: **Lewin**
- ☆ Term I: D Process : **Rogers**
- ☆ Concept of COIK fallacy : **Dale**
- ☆ Concept of ingroup and outgroup : **Semnar**
- ☆ Term social actualization : **Kurt Goldstein**
- ☆ Term ex post facto : **Chapin and greenwood**
- ☆ Comcept of employment : **David wener**
- ☆ Learning by doing concept : **John Dewey**
- ☆ Gemenischaft and Gesellschaft : **F Tonnies**
- ☆ Term Rurabinate : **Donald Tuttle**
- ☆ Sanskritization: **M N Srinivas**
- ☆ Theory of Laisserz faire : **Adam Smith**
- ☆ Satellite model of extension : **Perumal**
- ☆ Bond theory of learning: **E.L. Thorndike**
- ☆ Bio-village concept: **M.S. Swaminathan**
- ☆ Communication Competency: **Hymes**
- ☆ Word extension: **Vorhees**
- ☆ Concept of extension education process: **Leagans**
- ☆ Principles of adult learning: **Knowles**
- ☆ Bond theory of learning: **Thorndike**
- ☆ Cone of experience given by: **Edgar Dale**
- ☆ Cone of experience modified by: **Sheal**
- ☆ Steps in extension teaching presented by: **Wilson and Gallup**
- ☆ Father of PRA: **Robert Chamber**
- ☆ Farmer first model by: **Robert Chamber**
- ☆ Term andragogy: **Alexander Knapp**

- ☆ Credit for success of KVK: **Dr. Chandrika Prasad**
- ☆ Blue print for establishment of agri universities: **Dr. Ralph W. Cummings**
- ☆ Crop demonstration on cotton laid by: **Seaman A. Knapp**
- ☆ ICAR become autonomous body based on recommendation of: **Dr. Marison W Parker**
- ☆ IARI established under vice royalty of: **Lord Curzon**
- ☆ 3tier Panchayati Raj by: **Balawant Rai Mehta**
- ☆ 2tier Panchayati Raj by: **Ashok Mehta**
- ☆ Term Panchayati Raj: **J L Nehru**
- ☆ People's plan: **M N Roy**
- ☆ Term Hindu rate of growth: **Raj Krishna**
- ☆ Concept of Grameen Bank in Bangladesh: **Mohd. Younis**
- ☆ 4 H Club founded : **O H Benson**
- ☆ Motto of 4 H Club: **Carri Harrison**
- ☆ 4 H pledge worded : **Otis Hall**
- ☆ Young farmers association: **P S Deshmukh**
- ☆ Theory of mass communication: **Defleur**
- ☆ Role taking: **George Mead**
- ☆ Linguistic theory of communication: **williams**
- ☆ Balance theory of communication: **Heider**
- ☆ Medium is the message: **McLuhan**
- ☆ Communication IQ: **Bhanja**
- ☆ Communication competency: **Hymes**
- ☆ Key communicator concept: **Lazarsfeld**
- ☆ Dimensions of message: **Berlo**
- ☆ Message distortion: **Kirk**
- ☆ Concept of feedback: **D Berlo**
- ☆ COIK fallacy: **Edgar Dale**
- ☆ Term ID process: **Rogers**
- ☆ Adoption quotient: **Chattopadhyay**
- ☆ Adoption index: **Bose**
- ☆ Father of sociology: **August Comte**
- ☆ Father of modern sociology: **R K Meston**

- KVK: **Mohan Singh Mehta**
- Term primary group: **Cooley**
- Term Sanskritization: **M N Srinivas**
- Gemeinschaft and Gesellschaft: **Tonnie**
- Folkway: **Sumner**
- Social institution: **Horton and Hunt**
- Concept of PRA: **Henry Ford**
- Father of PRA: **Robert Chambers**
- Father of PRA in India: **Neela Mukherjee**
- Experiential learning cycle: **James A Mc Caffery**
- Functions of management: **Henri Fayol**
- Contingency theory of leadership: **Fielder**
- Path goal theory: **Robert House**
- Hierarchy of need theory: **Abraham Maslow**
- ERG theory: **Clayton Alderfer**
- Hygiene two factor theory: **Frederick Herzberg**
- Reinforcement theory: **B F Skinner**
- Goal setting theory: **Edwin locke**
- Expectation theory: **Victor H Vroom**
- Equity Theory: **J Stacy Adams**
- Theory X and Y : **Mc Gregor**
- Theory Z: **Oouchi**
- Management by organization: **Peter Druker**
- Delphi Technique: **Olaf Helmer**
- Managerial grid: **Robert Blake and Jane Mouton**
- Word entrepreneur: **Richard Cantillon**

IMPORTANT TERMS AND THEIR PLACES OF FIRST USE/ ORIGIN

Term/Concept*: *Country/Place

- Junior Farmers Club: **Australia**
- Ex workers as Rural agents: **Peru**
- Ex workers as agriculture Representatives: **Canada**
- Community development: **Cambridge (1948)**

- National Grange (Farmer Organization): **USA (1867)**
- Field School: **Indonesia**
- Massaggana 99 (TOT programme for Rice): **Philippines**
- Tukki Approach (multipurpose extension worker): **Nepal**
- Cruzada ABC (mass media programme): **Brazil**
- Camilla Project: **Bangladesh**
- Kibbutz: **Israel**
- Harvest Programme: **China (1987)**
- Telescuola (media forum): **Italy**
- Community based Landcare Programme: **Australia**
- Radio phonics: **Latin**
- School on Air: **Philippines**
- Media Forum: **Canada**
- Country (Extension Unit): **USA**
- BRAC: **Bangladesh**
- Cooperative Extension Service: **USA**
- Term 'PRA' first used in: **Kenya**
- Concept of Agro Ecosystem Analysis: **Thailand**
- Farmers First Model: **UK (University of Sussex) in late 80's**
- Commodity Focused Approach popular in: **Western Africa**
- Brainstorming originated in: **USA**
- Term 'Extension': **UK (1866)**
- Modern posters: **Paris**
- Flag method of extension teaching: **Latin America**
- Agro-ecosystem analysis: **Thailand**

FATHERS OF DIFFERENT DISCIPLINES

- Father of Learning: **Thorndike**
- Father of Demonstration: **Seaman A. Knapp**
- Father of Indian Extension: **K.N. Singh**
- Father of PRA in USA: **Seaman A Knapp**
- Father of Sociology: **Auguste Comte**
- Father of Modern Sociology: **R.K. Meston**
- Father of University Extension: **James Stuart**

- ☆ Father of Extension: **J.P. Leagans**
- ☆ Father of Diffusion Simulation: **Torsten**
- ☆ Father of PRA: **Robert Chamber**
- ☆ Father of PRA in India: **Neela Mukherjee**
- ☆ Father of Project management: **Henry Gantt**
- ☆ Father of Scientific Management: **Fredrick W. Taylor**
- ☆ Father of Modern Management theory: **Henry Fayol**
- ☆ Father of Androgogy: **Alexander Knapp**
- ☆ Father of Adult education in USA: **Lindmen**
- ☆ Father of Stress Management: **Hans Selye**
- ☆ Father of Social system approach: **Vilfredo Pareto**
- ☆ Father of Development administration: **George Gantt**
- ☆ Father of Modern Digital Communication: **Shannon**
- ☆ Father of Mass Communication: **Schramm**
- ☆ Father of India Telecom: **Sam Pitroda**
- ☆ Father of Wireless Communication: **J.C. Bose**
- ☆ Father of Brainstorming : **Osborn**

Chapter 2

Extension Concept and Scope

1. Education is an integral part of extension. So it is pursued to educate, motivate and change behavior of people.
2. Education is a PROCESS, *i.e.* the process of developing capabilities of individual so that they can adequately respond to their situation
3. COOMBS AND AHMED – 3 types of education (informal, formal, non-formal)
4. Target group specificity and fixed curriculum are the features of formal education
5. Similarity between formal and non-formal is that both are organized, dissimilarity is huge since one is institutionalized other is beyond the 4 walls of institution (refer to G.L. ray for more dissimilarity)
6. Six key element of non-formal education - Etling, Radhakrishna and Bowen [reliance on local resources, such that cost can be kept low, is a key element of non-formal education.(For other elements refer to G.L. ray)]
7. Concept of development – Dube

   1st phase: economic development (capital formation and trickle-down effect)

   2nd phase: institutional development (emphasis on radio and television)

   3rd phase: human centered development (panchayat body)

   4thphase: knowledge and competition

8. Main problem of developing nations in their lack of implementation
9. Concept of sustainability- Nicodemus
10. Sustainable demand – creation of political order in which firstly control of natural resources rests maximum extent possible with local community who are dependent on the resources. Decision making within community is as participatory, open and democratic
11. Objective of university extension is to take educational advantage of universities to the ordinary people
12. ENSMINGER: extension is education and that is possible to change attitude and practices of the people with whom the work is done
13. Leagans – extension education as an applied science
14. NCA (1976)- extension as an out of school education and services for the members of the farm family and others directly or indirectly engaged in farm production
15. Main aim of extension is HUMAN RESOURCE DEVELOPMENT
16. Extension maybe defined as a science of developing capabilities of people for sustainable improvement in their quality of life
17. Levels of extension
    a. Extension education (in which field work is generally limited)
    b. Extension service
18. Main responsibility of extension service is with state government
19. Extension service is – location specific, input intensive, target oriented
20. In university extension is integrated with teaching and research, whereas is research institute extension is integrated with only research
21. Main responsibility of extension service is educating and training farmers, farm women, rural youth and village leaders of the state
22. Philosophy – body of wisdom OR body of general principles or laws of a field knowledge OR view of life and its various components
23. Philosophy furnishes principles and principles/guidelines shape any program
24. Basic philosophy of extension is HOW TO THINK, NOT WHAT TO THINK (Kelsey and Hearne)
25. Sound extension philosophy is always forward looking
26. FUNDAMENTAL OBJECTIVE OF EXTENSION: develop the rural people economically, socially and culturally by means if education
27. FUNCTION OF EXTENSION: to bring about desirable change in human behavior by means of education

28. Changes can be bought by – knowledge, skill, attitude, understanding, goal, action, confidence
29. SKILL: technique of doing things, ACTION: performance of doing things
30. PROCESS: course of procedures, something that occurs is a series of action or event connecting to desired end
31. Effective learning experience is one which results in maximum no. of desirable changes in the behavior of learners
32. Learning should be experience centered *i.e.* farmers primarily learn by doing in addition to seeing and hearing
33. Drakenwald and Merriam – there is a change in time perspective as individual mature from one future application of knowledge to immediacy of application.
34. Attitude and values are the main reason for which adult learn than they might
35. Motivation is directly proportional to the rate of learning
36. Principle of learning: 11
    a. Principle of self-activity
    b. Principle of association [learning is growth like and continuous]
    c. Principle of dissociation [learning is motivated by emotions]
    d. Principle of transfer
    e. Principle of readiness
    f. Principle of set or attitude
    g. Principle of practice [learning must result in functional understanding]
    h. Principle of motivation
    i. Principle of timing
    j. Principle of clarity of objective [learning is purposeful]
    k. Principle of satisfyingnes

    Trick to remember is: MAD STRESS ? Practice Clarity Timing
37. Cone of experience provide linkage between – learning, activity and participant involvement
38. Cone of experience classify instructional methods on the basis of experiential correctness
39. Principle of extension: 11
    a. Principle of cultural difference
    b. Grassroots principle

   c. Principle of indigenous knowledge
   d. Principle of interest and needs
   e. Principle of learning by doing
   f. Principle of participation
   g. Principle of family
   h. Principle of leadership
   i. Principle of adaptability
   j. Principle of satisfaction
   k. Principle of evaluation

   Trick to remember: FILL CGP EASLI

40. Needs and interest of people are the starting point of extension work
41. Technology is the knowledge applied fir practical purposes
42. Agricultural technology have two components
    a. Material technology
    b. Knowledge based technology
43. Agro tech. is a complex blend of – (material+ process+ knowledge)
44. Any set of technology can be stated as valid tech. after it is systematically for its EFFECTIVENESS AND CONSISTENCY
45. When several alternatives of the technology are available then demonstrations are preceded by field trails mainly to prioritize the recommended practices based on field performances
46. Environment protection act 1986
47. Main purpose of EXTENSION WORK is to motivate people to adopt new ideas and practices, where extension agents acts as motivators
48. WILSON AND GALLUP'S MOTIVATION [S.E.R.R.]
    a. Desire for Security
    b. Desire for Experience
    c. Desire for Response
    d. Desire for Recognition
49. McClelland's motivation [P.A.A]
    a. Need for Power
    b. Need for Affiliation
    c. Need for Achievement

50. McClelland's need for Affiliation and W and G's desire for response is somewhat similar
51. Decision making is the process of consciously choosing courses of action from available alternative and integrating them for the purpose of achieving desired goal
52. Decision making influenced by
    a. Level of knowledge
    b. Cost involved
    c. Time availability
53. Learning to utilize intuition is of the crucial features decision making
54. Extension teaching is planned and deliberate act on the part of agent (extension agent)
55. Steps in extension teaching [AIDCAS] given by W and G
    a. Attention: starting point of change, mass method like Radio, TV, exhibition, mobile phone, and personal contact of extension agent and local leader are used in this
    b. Interest: furnish more information about the topic in a way that will be able to understand and use necessary to present one idea at a time. Here personal contact by extension agent, local leaders, farm publications are essential
    c. Desire: unfreezing existing behavior and motivating the people to change, emphasize on the advantage of new idea or practice. Here demonstrations, farm publication, personal contact by the extension agent and group discussion are essential.
    d. Conviction: Stage of strong persuasion, application of the new idea or practice in their own situation, furnishes adequate idea and how it works. Here field day, farmers day, slideshow, personal contact by the extension agent and training are essential
    e. Action: putting idea into practice, small scale demonstration. Here supply of the critical inputs essential in this stage.
    f. Satisfaction
56. In democratic group people participate based on ones skill and interest
57. Behavior has to be changed not by any force but by educating and motivating them
58. Quantitative evaluation – measuring changes as a result of extension program.
59. Qualitative evaluation – provide reason why extension agent farmer act in a certain way

60. 9 areas of program emphasis – Kelsey and Hearne
61. System is essential a set or assemblage of things interconnected or interdependent so as to form a complex unity
62. India has one of the lowest median ages
63. When an economy prospers first its death rate falls and then its birth rate falls
64. Capacity is building is crux
65. Most important challenge for future is management of knowledge sine in coming years success of a farmer will depend on his level of knowledge
66. So agriculture extension should focus on expanding decision making skills in the knowledge intensive way and nit just agricultural production alone.
67. Words;
    a. Vulgarization [French] – extension
    b. Voorlichting [Dutch] – lighting the past
    c. Al-ershad [Arabic] – guidance
    d. Berating [German] – Advisory body
    e. Capacitation [Spanish] – Capacity building
    f. Song suem [lao/Thai] – to promote
    g. Tarvij and goshtaresh [Persian] – to promote and to extend
    h. Penyuluhan [Indonesian] – give illumination in the dark
68. University extension
    a. Cambridge – 1873
    b. London – 1876
    c. Oxford – 1878
69. Dates:
    a. County commission 1862
    b. First agro society 1785
    c. Morill act 1862
    d. Smith liver act 1914
    e. Hatch act 1887
    f. AICRP on extension 1977
    g. ISEE 1964
    h. DEE 1966
    i. Extension [ICAR] 1971
    j. Agricensus [INDIA] 1970-71

70. No. of extension approaches according to Axinn – 8
71. Book " a guide to alternative extension approach" written by – G.H. AXINN
72. Project approach: the extension approach wherein short changes are measured of success *e.g.* – NAIP, IVLP
73. The extension approach wherein the success is measured by in terms of production increases of particular crops covered by the program: Training and Visit approach
74. The extension approach wherein success is measured by the extent to which farming people adopt the technologies developed by the program and continue using the overtime: Farming system development approach

    Also development of technology which fits the need is farming system approach
75. The extension approach wherein success is measured in terms of the rate of take up of the recommendation, and in national production – General agricultural extension approach
76. Four paradigms of extension education:
    a. Technology transfer – (persuasive + paternalistic)

       It involves top down approach that delivers specific recommendations to the farmers about the practices they should adopt.

    b. Advisory work – (persuasive + participatory)

       Participatory approaches by the govt. or NGO to promote, pre-determined packages of technology

    c. Human resource development: (educational + paternalistic)

       Dominated in earliest days of extension in Europe and USA

       Top down teaching method is adopted

    d. Facilitation of empowerment: (educational + participatory)

       Paradigm involves methods such experimental learning and farmer to farmer exchanges

       Knowledge is gained through interactive process and participants are encouraged to make their own decision ex. FFS (farmer field school)
77. Four generations of extension in Asia:
    a. Colonial agriculture
    b. Diverse top down approach
    c. Unified top down extension
    d. Diverse bottom up approach
78. Fundamental objective of extension education is making people to learn

Chapter 3

Extension Schemes

1. Main agency for agro development is ministry of agriculture at national level and dept. of agriculture at state level
2. Ministry of rural development embraces over 60 per cent of the agricultural activities
3. ICAR institutes and AUs play a major role in organizing demonstration training *etc.* on a limited scale and have a catalytic effect on other sub system
4. Divisions that are directly controlled by ICAR
   a. Technical divisions of crop soils, horticulture, agricultural engineering, animal science, fisheries *etc.*
   b. National research Centre
   c. Coordinated research projects
   d. Institute such as KVK
   e. Divisions of agricultural extensions – it controls zonal coordination units
5. Division that are directly controlled by AU
   a. Faculty/college of agriculture, horticulture, veterinary, dairy, agro engineering, forestry, home science *etc.*
   b. Directorate of research
   c. Directorate of extension – which control KVK

6. Divisions that are directly controlled by dept. of agriculture and cooperation
   a. Technical divisions for crops, horticulture, plant protection, fisheries
   b. Directorate of extension
7. 4 main TOT projects:
   a. AICPND 1964
   b. LLP 1979
   c. ORP 1974-75
   d. KVK 194-75
8. AICPND: 1964

   Rationale- unless the scientist could demonstrate what they have advocated their advice may not be heard by the farmer.

   No separate control plot and specific yield target.

   Farmers in whose plots the demonstrations are laid out are actual farmers with small holdings.

   Area of demonstration – 1 hectare (1 acre if bigger plot is not available)

   Demonstration was intended to show genetic production potentialities

   Objectives- to give training to both farmers and field extension functionaries to provide first-hand knowledge to research workers
9. ORP - 1974-75

   Aimed to disseminate proven technology in a discipline among farmers on a watershed basis

   Covering whole village or cluster of village

   Also studying the constraints to improve technical know-how

   Primarily devoted to demonstrate new technology on a large scale

   Attempted interdisciplinary and inter-institutional approach
10. KVK – 1974-75

    Also known as farm science Centre or agriculture science Centre

    It is actually need based skill oriented vocational training

    To practicing farmers in services field extension officers, those who wish to go for self-employment

    KVK will actually impart technical literacy

    3 fundamental principles of KVK
    a. Agricultural production as the prime goal
    b. Work experience are the main method of imparting training

c. Priority to the weaker section of the society

d. Organise vocational training.

Main idea was to influence productivity to achieve SOCIAL JUSTICE for the most needy and deserving weaker section of the society, No certificate are provided after the completion Of the training Main objective of KVK is to provide a strong training support for bringing about production breakthrough in agriculture Organizing farm science club is an objective of KVK and also to enhance a liking for agriculture

11. LLP – 1979

Initiated for 75000 farm families

Overall objectives: of the program was to improve economic condition of small and marginal farmers and landless agricultural laborers by TOT, developed by AU and research institute.

- Assist farmers in developing feasible farm plans
- Low cost relevant agriculture and allied tech in farm and home for increasing production and income
- Organize training program

12. REORGANIZED KVK – 1992

All the 1st line TOT of ICAR *viz.,* ND, ORP, LLP, integrated with KVK Major mandate of KVK

a. Training program

b. Frontline demonstration

c. On farm testing

For proper functioning KVK are to convene scientific advisory committee once in six months

To make KVK self-sustaining they are also encouraged to take up revolving fund scheme

A. Frontline Demonstrations: demonstrations conducted under the close of scientist of National Agricultural Research System (NARS)

Frontline demonstration coz. The technologies are demonstrated for the 1st time by scientist themselves before passing it to the main extension system

OBJECTIVE: to demonstrate newly released crop production and protection technologies and management practices in farmers field under different agro climatic conditions

B. OFT – main objective – problem diagnosis, planning, experimentation and extrapolation

13. University grant commission headed by S. Radhakrishnan [1949] – recommended establishment of rural universities
14. Two joint indo American team [1954-55] and [1959-60] recommended setting up of agro universities in the state in the pattern of land grant colleges in USA
15. Review of agricultural universities (M.S. RANDHAWA) 1978 – ONE OF THE ESSENTIAL FEATURES OF THE AGRI UNIVERSITY IS THE ACCEPTANCE OF PHILOSOPHY OF SERVICE TO AGRICULTURE AND RURAL COMMUNITY.
16. Extension educational institute [EEI] it is formed to cater the extension education and training needs of large numbers of extension personnel in the country

    It is primarily trainers training center

    EEI all over India

    a. Nilokheri (Haryana) – 1959
    b. Anand (Gujarat) – 1962
    c. Hyderabad (AP) – 1962
    d. Jorhat (Assam) - 1987
17. Mission of MANAGE: to facilitate the extension officers, managers, scientist and administrator in all sectors of agriculture economy to acquire managerial and technical skill that would help to them to facilitate farmers for practicing sustainable agriculture
18. FAMINE COMMISSIONS:
    a. 1860 (STRECHY) – recommended establishments of agricultural dept. under the tenure of Lord Mayo
    b. 1880 (Lyal) – revival of department of agriculture under the control of secretary (1881), and all provinces o have dept. of agriculture in 1882
    c. 1901 (McDonalland) – setting up of IARI, appointment of experts capable of applying scientist methods to do improvements in agriculture
19. Royal commission on agriculture (1928) – said that- research can be of no help to the cultivators until the results are given to them in a form in which they become a part of their agro produce.

    Appointment of a director of agriculture whose sole duty would be to organize and systematize propaganda work throughout the province
20. 1st government sponsored scheme – Firka development scheme 1946.

21. Gurgaon project is the first program initiated on a large scale. F.L. brynn was deputy commissioner of Gurgaon district Gurgaon experiment was based on the principle of rural development on practical basis Here propaganda was done by drama and music
22. Sevagram project was actually a gandhian constructive program. Concept: helping people to help themselves Basic education was through "learning by doing and earning while learning" The single cause of the failure of gandhian program is -industrialization in the nation
23. Marthandam project Dr. spencer hatch was an agricultural expert, he used agricultural development as a bait to convert Hindus into Christianity Earlier it was a 3 field program – development of spirit, mind and body. But later it developed into 5 fold program psychic, spirit, mind, economic and social aspect 6 weeks short training courses to villagers and school teachers were also done.
24. Firka vikas yojana 1946 Attainment of gandhian ideal of "GRAM SWARAJ", it was launched in the last quarter of 1946 in 34 firkas. It was later extended to 50 more firkas so total firkas become 84. This is around 2 firkas per district. Each firka is divided into 5-10 villages.
25. Etawah pilot project 1948 IT HAS SHOWN ITS WAY TO THE COMMUNITY DEVELOPMENT IN INDIA, that's why it called Etawah pilot project,. Forerunner of CDP Started in 64 villages in mahewa UP Introduction of multipurpose VLW Village level worker – 4-6 villages Group level worker – 1 block (100 villages) District planning officer – 1 district
26. Nilokheri 1947 Name of the project is "majdoor manzil" It had the objective of making 700 acres of swampy land cultivable Polytechnic training for BDO, SEO and VLW
27. Fiscal commission 1949 Setting up an extension service in India with objective of bridging the gap between research and products Found that all of the dept. are working under water tight compartments GMF (1st time in 1943 again after independence in 1947) Additional staffs were provided in district and taluk level GMF enquiry committee 1950- headed by V.T. Krishnamachari Pattern of staffing for a tehsil or taluka

    a. 1 BDO
    b. 4 technical officers
    c. 12 VLW

    Based on the feedback of the GMF enquiry committee report and etawah pilot project experiment planned to start 15 pilot projects with financial assistance from the FORD FOUNDATION
28. Indo – US technical cooperation agreement was signed in January 1952, initiating 55 CDP in 27388 villages of population 16.4 million for a period of 3 years

29. Each project – operational area – 400-500 sq. miles Comprising of – 300 villages Population – 2 lakhs Divided into – 3 developmental blocks consists of 100 villages VLW no. – 60/each division, so for each VLW there is 5-10 villages.

    Headed by – project officer

30. 4 stages of CDP:
    a. Conception (3 months) – selecting of area, economic survey, planning
    b. Initiation (6 months) – arrangement of temporary houses and other things necessary for operation
    c. Operation (18 months)
    d. Consolidation (6 months)
    e. Finalization (3 months)
31. Entire country covered by CDP in 1958 and new CDP has self life of 10 years
32. CDP was transferred to state sector in 1959
33. NES block was headed by BDO, comprising of 100 villages, 65 thousand population and 10 VLW and at an area of 150-170 sq miles
34. NES is an agency and CDP is a method
35. Extension officers are under the technical control of respective district level officers technical program
36. BDO is the entrepreneur and the key person in CDP
37. VLW are the friend philosopher and guide of village farmers
38. Community development originated in Cambridge
39. 1948 summer conference on African administration recommended this term as a movement
40. Taylor *et al.,* describes community development as a method
41. Mookherjee said it as a 1. Method 2. Program 3. Movement
42. Steps in community development process – Dwarikanth 1967
43. Critical analysis of India's CDP is done by – Taylor et. Al
44. Team for the study of community development and national extension service is headed by Balvantray g. Mehta (who again recommended democratic decentralization)
45. Difference between community development and extension education
    a. Extension education is a more indirect approach, to help people to help themselves. Whereas Community development is a direct govt. approach

b. Extension education aims to bring change by leaving decision for change of individuals, whereas in CDP decision are taken by group or representatives

c. Extension education is a mono purpose approach, whereas CDP is a multipurpose approach

d. Extension education is a MEANS. Whereas CDP is the END

46. 3 phases of panchayati raj

    a. Phase of ascendancy 1959-1964

    b. Phase of stagnation 1965-1969

    c. Phases of decline 1969-1977

47. 73rd amendment act 1992, ministry of panchayati raj 2004

    *** For the rest of panchayti raj kindly follow ANGRAU notes, and also from other sources

48. "food enough should be a central objective in the crusade of new India" – 1st agricultural production team of ford foundation

49. Idea behind IADP – increased agricultural productivity shall lead to economic growth which again shall lead to welfare of the society

50. IADP started in 7 selected districts later 21 more districts are added Implementation of IADP – 1st year 20 per cent area and the end of 5 years 65 per cent area Composite demonstration instead of single demonstration Early maturity and photo insensitivity in most cases provide flexibility in their period of planning which facilitates multiple cropping – HYVP

51. T&V system also known as BENOR SYSTEM [KHARIF 1974] Article "agriculture extension training and visit system" written by Daniel Benor and James Q Harrison T&V approved by govt. of India in 6th planning period World Bank assistance program The purpose of T&V system was to build a professional extension which will guide farmers to increase the yield The OBJECTIVE of T&V system is to remove the shortcoming in the existing agricultural system Basic spirit behind T&V program is – even though farmers haven't produce any satisfactory crop in the past but still it can produce good crop within an agricultural season provided that he have to do to what he will be advised by the guider

    Salient features of T&V system

    a. Professionalism

    b. Single line of command

    c. Time bound work

    d. Field and farmer orientation

e. Regular and continuous training
f. Linkages with research

Transfer of know-how from the SMS to the famer was ensured in stages

a. Training: for the transfer of know-how from SMS to extension worker
b. B. visit: for the transfer of knowhow obtained at the training from extension worker to the farmer

Organization of the T&V system:

1 AEO SUPERVISE 6-8 SDEO, SDEO supported by SMS team.

4-8 SDEO supervised by DEO, who is again supported by SMS

LEVEL EXTENSION PERSONNEL

COVERAGE

Field level village ext. officer (VEW) 800-1200 farmers Block level agri. Ext. officer (AEO) 8 VLW/VEW Sub divisional SDEO 6-8 AEO District level DEO 4-8 DEO in districts 800-1200 farmersare divided into 8 groups of equal size and from them 10 per cent of the farmers are selected as contact farmers.

TIME BIFURCATION OF VEW, AEO and SMS

Now there are 8 groups of farmers, so VEW have to give 8 days to the farmers such that it can cover one day per group. So VEW spend 4 days/ week, so in a fortnight (14days) it will become 8 days. Now remaining 2 days/week (since there are 6 working days per week), 1 day is dedicated to AEO for in-service training, and the other one is for unscheduled visit In a fortnight there is 1 training given by SMSto the VEW And weekly training is conducted by AEO, which we discussed earlier AEO spend 2 days/fortnight in conducting training to the VEW, and spend 8 days/ fortnight in field supervising the VEW, now for the remaining 2 days/ fortnight in takes training from the SDEO. SMS spends 1/3rd of their time to VEW, 1/3rd of their time in the field and 1/3rd of their time in visiting research facilities. Groot analyzed the limitation of T&V system

- ☆ NAEP (1983) was came to overcome the constraints of T&V system, in which basic objective is the bridge the gap between research and extension
- ☆ Technology mission-is the special assignment given of task given to a group of people to fulfill a technological program within a given period of time; a technology mission is set up when a critical need to boost up certain technology arises.
- ☆ TMO (1986) – it is for edible + non-edible oil

52. IRDP 1979

Introduced in 1978-79 in 2300 blocks of the country and after some time in 1980 it was introduced I n all the blocks of the country Each block will get 5 lakh rupees or 3000 families in every block will get direct help within 5 years. In every block 600 families are selected. Among total selected families 50 per cent of them will be the families of SC and ST. 40 per cent of them women and 3 per cent persons are disabled and remaining 7 per cent are general poor families Management is guided by DRDA, the ratio of funds shared by center and state is 50:50 respectively

Pattern of assistance of subsidy:

a. Small farmers – 25 per cent
b. Marginal farmers – 33.33 per cent
c. SC and ST – 50 per cent

Basic approach of IRDP – family approach

Basic unit development - CD block

53. JRY [1989] = NREP + RLEGP

Fund allocation pattern between center and state is 80:20 respectively At least 80 per cent of the fund of the allocation under the program transferred to the village panchyayat Fund distribution pattern among zilla parishad, panchayat samiti and village panchayat is 20:5:75 respectively This was implemented all over the country via village panchayat 30 per cent of the employment opportunities under the yojana are reserved for women Primary objective was to give additional gainful employment to the rural people whereas secondary objective was to build community assets for the rural poor

** 1985-86 IAY was started as a sub scheme of RLEGP, but when RLEGP merge with JRY, IAY was also a part of JRY. But later in 1st January 1996 IAY was separated from JRY

54. EAS 1993

Employment for 100 days in lean agricultural season Scheme was implemented in the backward block of the district, situated in the drought prone areas, deserts areas tribal areas and hilly areas Weightage under EAS – 40 per cent greenery work and 20 per cent building work All work under EAS under labor intensive only

55. DWCRA 1982-83

Sub scheme of IRDP, launched in pilot basis in 50 districts, covered all district in 1990 TRYSEM fulfill the skill requirement of DWCRA Approach:

10-15 women form a group one member of the group was the organizer, who assisted in the choice of activity procurement of raw material marketing of produce Eligibility: rural women coming under the BPL have to form self-help group. 50 per cent of the women must belong to SC and ST. priority was given to physically handicapped persons and girls or women

56. SGSY 1999

    Major self-employment program Program to aimed to creating self-employment opportunities for rural poor with emphasis on SHGs by providing them subsidy and bank credit

    TYPE SUBSIDY MONEY SUM

    Normal individual 30 per cent Rs. 7500 PH/SC/ST 50 per cent Rs. 10000 SHG 50 per cent Rs. 125000 Aims to establish a large no of microenterprises in the rural areas beneficiary known as swarojgaris may be individual or groups Objective is to bring every family above the poverty level in 3 years Key activities are selected with the approval of panchayat samiti at the block level and DRDA at the district level

57. NATP

    Under NATP innovation in technology dissemination was planned ATMA objective: integrating research, extension and all other stakeholders at the district level of support the farmer's needs and interest through integrated approach ATMA is a society of key stakeholders and is bottom up in nature ATMA is a registered society ATMA is managed by project director at district level Initially ATMA launched in 28 districts of 7 states ATMA organizational structure:

    a. ATMA governing body(advisory body) – 16 members, DM is the head of advisory body
    b. ATMA management committee (operational body) – 12 members, project director is head FIAC work under ATMA at block level SAMETI-apex training institute at state level to give training support to ATMA. SREP is developed by ATMA

58. Under ministry of rural development comes

    a. NIRD
    b. CAPART
    c. MGNREGA

59. Highest policy making body in matters related to national planning is – national development council (NDC)

PRE-INDEPENDENCE EXTENSION AND RURAL DEVELOPMENT PROGRAMMES

Introduction

About 65 percent population of the country is living in the economically undeveloped area. Therefore, many evils such as illiteracy, superstitions, unhealthy practices *etc.* took roots in our country. After Second World War or after independence, it was thought to settle all these problems properly and so, many development programmes were reinvigorated. It is wrong to declare that there was nothing done during slavery period. However, those works, which were executed at that time, were very few in number according to the needs. Looking at a vast country like India, during British rule some selected social workers had started some programmes of rural development. The evolution of extension programme is described in two stages:

Stage I- Pre-Independence Era (1866-1947)

Stage II- Post-Independence Era (1947-1953)

1. Stage I. Pre-Independence Era (1866-1947)

During the pre-independence era, various attempts were made by the individual persons and private agencies *viz.,* Gurgaon project in Haryana by Mr. F.L. Brayne; Sriniketan Experiment in Bengal by Sri Rabindra Nath Tagore; Servants of India Society, Pune, in Tamil Nadu, Central India (presently Uttar Pradesh and Madhya Pradesh); Sir Daniel Hamilton s Scheme of Rural Reconstruction at the Sunderbans in Bengal; Rural Reconstruction Work by Christian Missions through Allahabad Agricultural Institute; The Christian College, Nagpur; Marthandam project by Dr. Spencer Hatch under the auspices of YMCA; Sevagram experiment in Wardha by Mahatma Gandhi; Adarsh Seva Sangh, Pohri, Gwalior; Indian Village Service in Lucknow and Etawah in Uttar Pradesh by Mr. A.T. Mosher and Sarvoday Programme in Bombay province by Sri B. N. Gupta.

It was during this period that the Department of Agriculture came into being in June 1871 under the then Government of India, and by 1882, agricultural departments in most of the provinces started functioning in skeleton form. Recognising the need for new and improved methods of cultivation based on agricultural research, the then Government of India also set up an Institute of Agricultural Research at Pusa in Bihar in 1905.

Rural Development Programmes

Year	*Place*	*Person/Agency*
1903	Scheme of Rural Reconstruction at the Sunderbans in Bengal	Sir Daniel Hamilton
1920	Gurgaon project in Haryana	Mr. F.L. Brayne
1920	Sriniketan experiment in Bengal	Sri Rabindra Nath Tagore

Year	*Place*	*Person/Agency*
1920	Seva-gram experiment in Wardha in Gujarat	Shree Mahatma Gandhi
1928	Marthandam project in Kerala	Dr. Spencer HatchYoung Men Christian Association (YMCA)
1932	Baroda village reconstruction project	Shree B.T. Krishnamachari
1942	Grow more food campaign	Agricultural departments of state governments
1945	Indian village service	Dr. W.H. Wisher
1946	Firka Vikas Yojana in Madras (now in Tamil Nadu)	Government of Madras

1.1 Scheme of Rural Reconstruction at the Sunderbans in Bengal

In 1903, Sir Daniel Hamilton formed a scheme to create model villages, in an area in Sunderban (Bengal), based on cooperative principles. He organised one village of this type and set on Cooperative Credit Society which functioned up to 1916. In the 1924, he organised a Central Cooperative Bank and Cooperative Marketing Society and established a Rural Reconstruction Institute in 1934.The institute provided training facilities in cottage and subsidiary industries.

1.2 Gurgaon project (1920)

In Gurgaon district, this programme of village development was the first one to be run by the State. It was started by the Mr. F. L. Brayne. In 1920, Mr. Brayne had been appointed on the post of Deputy Commissioner in Gurgaon district and he began this project of rural upliftment in his district, which became famous as Gurgaon Project . The main objectives of this project were: (1) To increase crop production, (2) to control extra expenditure, (3) to improve the health, (4) to develop the feeling of women-education, and (5) home development work. Although this project got some success yet this scheme also could not survive more because this project was also based upon the sentiments of F.L. Brayne and when he was transferred, gradually this programme also stopped.

1.3 Shriniketan Project (1920)

Shriniketan is situated about 100 km. away from Calcutta (Kolkata) in West-Bengal State. This area was backward socially, economically and politically. Shri Ravindra Nath Tagore began this project of village development in this area with the help of sociologist Shri L.M. Hurst. Shri Tagore thought that if some villages were developed, the other villages will get inspiration and the programme of village development will spread all over the country and thus the whole country would be developed. Shriniketan Village Welfare Institute was established in 1920 whose main objectives were the following:

1. To increase the knowledge of rural people.

2. To help the rural people in establishing cottage industry.
3. To inspire the people to follow new technology.
4. Development of dairy farming.
5. To create the feeling of co-operation.
6. To arrange the facility of health and education.
7. To create the feeling of rural leadership.

This Project, in the beginning, was started only in 8 villages but after some time this project was extended to 15 villages. Shri Tagore was the landlord of all these villages. Being centralized over Shri Tagore this Project could not survive after the death of Shri Tagore, and came to an end after his death.

1.4 Seva-gram (1920)

Mahatma Gandhi was a great social worker. He knew very well that as long as Indian people are suppressed, their society and their nation cannot progress. For ending this suppression, he began this welfare project SEVA GRAM , establishing his Ashram in wardha, in 1920. The main objective of this programme was to prevent the economic and social suppression of the people and to create the feeling of patriotism among them and they must think that this is their own country. For fulfilling this objective, Gandhiji made a programme, which became famous as Gandhian Constructive Programme. The main objectives of this project were: (1) to use khadi clothes, (2) to introduce Health Programme, (3) the programme of sanitation in the village, (4) the programme of women welfare, (5) the programme of economic help, (6) to uplift the backward classes, (7) primary and adult educational programmes, (8) to improve the conditions of poor people, (9) programme of social harmony, and (10) to popularise the mother tongue and other national dialects.

For making his programme successful and effective, Gandhiji had established All India Village Industry Association, All India Spinners Association, Hindustani Education Association and Kasturba Gandhi Association *etc.* Gandhiji s constructive programme was not followed completely by majority because those things, which were made by machines, were more beautiful and low-priced, and the common people could not be impressed. The single cause of failure of Gandhiji s programme was industrialization in the country.

1.5 Marthandam Project (1928)

This project of rural development was initiated in the village Marthandam near Trivendrum of Kerala State by Young Men Christian Association (YMCA) and Christian Church Association under the direction of Dr. Spencer Hatch in 1928. This village was undeveloped economically and the economic condition of the native majority was poor. Here, people used to cultivate only paddy and coconut on some places. For exploiting this weakness, it was thought that some developmental work should be done, so that the Christian faith could spread. Consequently Dr. Hatch made agreement with YMCA and the Church for his work and initiated this project in neighbouring village Marthandam.

The objectives of this project were also made on the basis as to how the public of this area could become Christian. The main objectives of this project were:

(a) Spiritual development

(b) Mental development

(c) Physical development

(d) Social development

(e) Economic development

This programme of Dr. Hatch remained for a long period because he had trained workers and the local people were inspired to work themselves. As such this project was very successful in this area. But the whole programme was centred on a person and after the death of Dr. Hatch, there was a conflict on the question of leadership between YMCA and Church Association and due to no economic advantage to the non-paid workers, this programme could not survive.

1.6 Baroda Village Reconstruction Project (1932)

Shree B.T. Krishnamachari in Baroda in the Gujarat State initiated this Project in 1932. This Project Continued for a long period.

a) *Objectives*

To improve the life style of rural people rapidly.

To spread education and industrialization.

To develop the necessary factors for the progress of agriculture.

b) *Scope*

This Project was started in the district of Navsari in the Gujarat State.

Many programmes such as gardening, poultry-farming, bee-keeping, spinning and weaving were organized.

Re-stabilization of Panchayats and other programmes of village progress were organized.

The adult education had been extended.

c) *Working System*

By personal education and contact by the village guides.

To use the school teacher of village in the extension of programme.

To use the traditional means of extension.

1.7 Grow more Food Campaign (1942)

This campaign, started in 1942, was continued after getting the independence. The main object of this campaign was to fulfil the need of food, which had been created due to the Second World War. This campaign was the first one to be organized on a national level. In this campaign, the new seeds and chemical fertilizers were distributed among the farmers. Agricultural departments of state governments organized this campaign.

a) Achievements

After spending a lot of money, some problems were tackled in the end.

The crop production increased due to new seeds and chemical fertilizers provided to the farmers.

From this campaign, the production of cotton and jute increased appropriately.

b) Limitations

Its work field was limited.

The campaign was deemed to be temporary hence the work was done.

The economic help, the distribution of good seeds and fertilizers could be available only on some places.

1.8 Indian Village Service (IVS-1945)

In 1945, under the guidance of Dr. W.H. Wisher, this service began in the village Agasoli, District Aligarh in U.P. But, after the partition of the country, the village volunteer M.V. Siddiqi Khan went to Pakistan and this centre was closed.

1.9 Firka Vikas Yojana (1946)

The government of Madras (now Tamil Nadu) decided to make efforts for the development of villages at Firka level. The first programme began in 1946. Among Pre-Independence project, this was the biggest project.

POST-INDEPENDENCE EXTENSION AND RURAL DEVELOPMENT PROGRAMMES

1. Introduction

The urgent need for stepping up food production was realised even in the pre-Independence era and a Grow-More-Food Campaign was started. Under the campaign, targets for increased agricultural production were laid down for the first time on an all-India basis. But the campaign failed to achieve its targets. Soon after Independence (1947), the Central Government re-defined the objectives of the Grow-More-Food Campaign as the attainment of self-sufficiency in food grains by 1952, and simultaneously increased the targets of production of other crops to meet the shortfall as a result of the partition of the country. At the same time, arrangements were made for integration and co-ordination of the entire campaign for increasing agricultural production. Some state governments associated the public with working of the campaign by setting up of non-official committees at the village, *taluka*, district and state levels. The plans were revised from time to time in order to make the campaign more effective. Accordingly, some initiatives in this direction were taken, which are given as under:

1.1 Etawah Pilot Project (1948)

The idea of intensive all-round development work in a compact area was put into practice as a Pilot Project in Rural Planning and Development in the Etawah District in Uttar Pradesh in 1948, which can be regarded as a forerunner of the Community Development Project in India. Albert Mayer, an American Engineer, played the key role in the initiation and implementation of the project. The programme was based on the principle of self-help, democracy, integrated approach, felt needs of the people, rigorous planning and realistic targets, institutional approach, co-operation between governmental and non-governmental organizations, close co-ordination between the extension service and the supply agencies and the collaboration by technical and social scientists. After an initial period of trial and error, a new administrative pattern was evolved. It percolated to the village level; the activities of different nation-building departments were channelled through one common agency and a multipurpose concept of village level worker was introduced. Each village level worker looked after 4-5 villages. The project was supervised by a district planning officer assisted by four specialist officers and other supporting staff.

Community Development

Year	*Programme/Scheme*	*Person/Agency/Approach*
1948	Nilokheri experiment (Mazdoor Manzil)	Shree S. K. Dey
1948	Etawah pilot project	Albert Mayer
1952	CDP	Community Development Programme
1953	NES	National Extension Service
1954	CDB	Community Development Block
1957	Panchayati Raj	Panchayati Raj Democratic Decentralization

Technological Development

Year	*Programme/Scheme*	*Person/Agency/Approach*
1952	Key village Scheme	Government of India
1960	IADP	Intensive Agricultural District Programme
1964	IAAP	Intensive Agricultural Area Programme
1963-64	ICDP	Intensive Cattle Development Project

Year	*Programme/Scheme*	*Person/Agency/Approach*
1966	HYVP	High Yielding Variety Programme
1974	T&V	Training and Visit System
1979	LLP	Lab to Land Programme
1983	NAEP	National Agricultural Extension Project
1986	TMO	Technology Mission on Oilseeds
1998	NATP	National Agricultural Technology Project
1999	TAR-IVLP	Technology Assessment and Refinement Institute Village Linkage Programme
2000	ATIC	Agricultural Technology Information Centre
2006	NAIP	National Agricultural Innovation Project

Other Developmental Programmes

Year	*Programme/Scheme*	*Person/Agency/Approach*
1970-1971	SFDA	Small Farmers Development Agency
1970-1971	MFAL	Marginal Farmers and Agricultural Labourers Programme
1970-1971	DPAP	Drought Prone Areas Programme
1972-1973	PPTD	Pilot Project for Tribal Development
1974	T&V	Training and Visit Programme
1978-1979	IRDP	Integrated Rural Development Programme
1979	TRYSEM	Training of Rural Youth for Self-Employment
1980	NREP	National Rural Employment Programme
1982	DWCRA	Development of Women and Children in Rural Areas
1989	JRY	Jawahar Rojgar Yojana

1.2 Nilokheri Experiment (1948)

Shree S. K. Dey, was the founder of the Nilokheri Experiment, when nearly 7000 displaced persons from Pakistan and later integrated with the 100 surrounding villagesin to what came to be rural-cum-urban township. It was built round the vocational training centrer Nilokheri, in July 1948. Shree Dey began this project using 1100 acre of swampy land spreading in the midst of Karnal and Kurukshetra. Shree Dey went on to become the Union Minister of Community Development in 1965. The scheme was also known as "Mazdoor Manzil" because of its principle 'he who will not work, neither shall he eat'.Under this programme, the main activities were a Vocational Training Centre run on co-operative lines and the Colony had its own dairy, poultry, piggery, printing press, engineering workshops, tannery and bone-meal factory. People were given vocational training of their choice to run these Co-operative Enterprises. Rights for education and medical care for the sick were also guaranteed.

1.3 Community Development Project (1952)

As a result of the Grow-More-Food Enquiry Committee Report and the successful experience of the Etawah Project, 15 Pilot Projects were started in 1952 in selected states with the financial assistance received from the Ford Foundation. Besides helping to increase agriculture production and bettering the overall economic condition of the farmers, these projects were meant to serve as a training ground for the extension personnel. It was soon realised that for the creation of an urge among the rural population to live a better life and to achieve permanent plenitude and economic freedom in the villages, a much bolder and dynamic effort was called for. It was recognized that the success of this new effort depended upon and whole hearted co-operation of the beneficiaries, government officials and non-officials at every stage, the education of rural masses in the technique of rural development and the timely provision of adequate supplies of the needed inputs and other requirements. For undertaking this new programme, the Government of India entered into an operational agreement with the Government of the USA under the Technical Co-operation Programme Agreement. Under this Agreement, 55 Community Development Projects were started in different parts of the country on 2nd October, 1952 for three years. The Projects covered nearly 25,260 villages and a population of 6.4 millions. Each project, in turn, consisted of about 300 villages covering 400-500 square miles and having a population of about two lakhs. The project area was divided into three development blocks, each comprising 100 villages and a population of 60,000 to 70,000. The development blocks, in turn, were divided into groups of 5-10 villages, each group being in the charge of a multipurpose village-level worker. The main aims of these projects were: to increase agricultural production by all possible means, to tackle the problems of unemployment, to improve village communications, to foster primary education, public health and recreation, to improve housing, to promote indigenous handicrafts and small-scale industries and to improve the villager's lot through their own primary effort. In short, the programme aimed at achieving all-round socio-economic transformation of the rural people.

1.4 Grow-more-Food Inquiry Committee Report (1952)

Though efforts were made to revitalize the Grow-More-Food Campaign, it was observed that the system was not functioning properly and the cultivator's response to the programme was very poor. As a result, the Government of India in 1952 appointed a committee known as the GMF Enquiry Committee to examine the working of the Grow-More-Food Campaign. The findings of this Committee revealed that the problem of food production was much wider than the mere elimination of food imports and that agricultural improvement was a very important part of a much wider problem of raising the level of rural life in the country. The Committee came to the conclusion that it was only by bringing about an appreciable improvement in the standards of rural life to make it fuller and richer that the rural masses could be awakened to take interest in not only increasing agricultural production but also improving their own conditions and creating a will to live better. The committee also pointed out that:

- ✰ All aspects of village life were interrelated,
- ✰ Improvement could be brought about by a number of detached programmes operating independently,
- ✰ There was lack of unity of efforts,
- ✰ The available finances was not adequate

The rural community as a whole did not participate effectively in the campaign. In short, *the movement did not arouse nation-wide enthusiasm and did not become a mass movement for raising the level of village life.* In its recommendations, the Committee proposed the formation of development block, each consisting of 100 to 120 villages, and the appointment of revenue officers as development officers or extension officers, assisted by technical officers for agriculture, animal husbandry, co-ordination and engineering. For actual work in villages, the Committee suggested the appointment of one village level worker for every five or ten villages. The Committee also described broadly the functions of the extension service, the manner in which the extension organisation would operate, the arrangements required in training of the required staff, the way in which the non-official leadership should be associated with the work of village development at the village, *taluka*, district and state levels. The need for setting up an independent organisation on the suggestions made, the manner in which the assistance should be rendered to the state governments as well as to villagers for development work, the role of the central and state governments in this effort was also emphasized by the committee. Based on these recommendations, the Planning Commission, which was set up earlier by the Government of India to prepare a plan for development consistent with the available resources, gave the highest priority to the development of agriculture and irrigation in the First Five-Year Plan. The Commission fixed substantially high targets of internal production and decided, as recommended by the Enquiry Committee, that the drive for food production should form part of plans for overall agricultural development, and that agricultural improvement in its turn should form an integral part of the much wider efforts for raising the level of rural life. The Commission prescribed "Community Development" as the method for initiating the process of transformation of the social and economic life of villages and "Rural Extension" as its agency.

1.5 National Extension Service (1953)

The National Extension Service was inaugurated on 2nd October 1953. The scheme of National extension service was designed to provide the essential basic staff and a small fund for the people to start the development work essentially on the basis of self-help. The operational unit of this service was an N.E.S. block comprising about 100 villages and 60,000 to 70,000 people. The N.E.S. blocks were later converted into community development blocks which had higher budget provisions in order to take up more intensive development programmes. The pattern of community development programme was further revised (modified with effect from 1st April 1958). According to this pattern, there were four stages:

- Pre-Extension Stage
- Stage I Blocks
- Stage II Blocks
- Post-Stage II Blocks

In this new set-up of community development blocks, all the nation-building government departments were brought together; and in order to ensure co-ordination at the block level, a new post of a Block Development Officer (BDO) was created. This officer is the co-ordinator of the programme and team leader and is supported by 8 extension officers drawn from the development departments, one each from the fields of agriculture, animal husbandry, co-operative, panchayat, rural industry, rural development, social education and welfare of women and children. Each normal block was provided with 10 village level workers and two *gram sevikas* (lady VLWs). Under this new set-up, the block is treated as an administrative unit for all the development departments, and the village-level worker is the contact person between these departments and the people.

1.6 Key Village Scheme (1952)

Key Village Scheme launched in August 1952 was the first systematic attempt to improve the quality and productivity of cattle and buffaloes in the country. It was proposed with a view to promote intensive cattle development in compact areas by utilising the limited quantity of high quality breeding stock. Basic objective of the scheme is the rapid multiplication of crossbred cattle for meeting the acute shortage of high quality animals by grading up of the indigenous cattle population. Gradually it embraced all the major aspects of cattle development such as superior breeding, castration of scrub bulls, fodder development, control of diseases, maintenance of records, milk recording and marketing of livestock and livestock products.

1.7 Intensive Agricultural District Programme (1960)

With a focusing on increasing productivity and production IADP was launched in July 1960 in 7 selected districts of various states in its first phase and later extended to another 9 districts more in second phase during 1963-64. This programme was aimed at integrated and intensive approach to solve the problems of agricultural production through adoption of package of improved practices. The achievements accomplished through this programme were tremendous and it really made a dent into the rural poverty for its eradication through agriculture. This programme was popularly known as Package Programme.

1.8 Intensive Cattle Development Project (1963-64)

The initiation of Key Village Scheme in August 1952 was the first systematic attempt to improve the quality and productivity of cattle and buffaloes in the country. The Intensive Cattle Development Project (ICDP) was initiated in 1963-64. The objectives of the project were the same as the Key Village Scheme, but had much wider in scope. It envisaged provision of a package of improved practices to the cattle owners to effect a breakthrough in milk production (National Commission of Agriculture, 1976). The package included

- Cattle breeding
- Castration of scrub bulls
- Veterinary aid and disease control
- Registration of milk recording
- Introduction of high yielding milch cattle
- Subsidies and incentives for milk production
- Feed and fodder development including demonstration, distribution of fodder seeds, silage making and popularization of chaff cutters and
- Dairy extension

The programme was launched in the milk shed area of major dairy plants to ensure the increased and steady flow of milk to these plants. There is no doubt that ICDP did help in increasing milk production in milch animals in the country to a great extent.

1.9 Operation Flood (1970)

Operation flood (also known as the White revolution of India) was a rural development programme started by India s National Dairy Development Board (NDDB) in 1970. One of the largest of its kind, the programme objective was to create a nationwide milk grid. It resulted in making India the largest producer of milk and milk products, and hence is also called the white revolution of India. It also helped reduce malpractices by milk traders and merchants. This revolution followed the Indian green revolution and helped in alleviating poverty and famine levels from dangerous proportions in India during the era. Dr. Varghese Kurien is known as Father of White Revolution.

The Operation Flood was implemented in three phases.

1. Phase I (1970 1980)
2. Phase II (1981 1985)
3. Phase III (1985 1996)

1.10 Training and Visit (T&V) System (1974)

The Training and Visit (T&V) system, the brain child of Dr. Daniel Benor (World

Bank Consultant) was introduced in India in 1974 for all round development of agricultural extension system in the country to introduce observation, training and technology transfer to the farmers and extension workers so as to enable them achieving greater productivity and production in the agricultural sector. This system has been introduced at first in 1974 in two states *viz.,* Rajasthan and West Bengal and then spread to another 16 states of the country aimed at building a professional extension service that is capable of assisting farmers in raising production, increasing incomes and of providing appropriate support for agricultural development. The important features of this system include professionalism, as single line of command, concentration of effort, time bound work, field and farmer orientation, regular and continuous training and close linkages with research. The creation of a dynamic link between farmers, professional extension workers, and researchers was done through the training and visit systems of extension. By adopting and implementing T&V system in its extension network India gained a lot in increasing its food production with scientific means coupled with effective technology transfer system.

1.11 Integrated Rural Development Programme (1978)

The Integrated Rural Development Programme (IRDP) is a rural development program of the Government of India launched in Financial Year 1978 and extended throughout India by 1980. It is a self-employment program intended to raise the income-generation capacity of target groups among the poor. The target group consists largely of small and marginal farmers, agricultural labourers and rural artisans living below the poverty line. The pattern of subsidy is 25 per cent for small farmers, 33per cent for marginal farmers, agricultural labourers and rural artisans and 50 per cent for Scheduled Castes/Scheduled Tribes families and physically handicapped persons. The objective of IRDP is to enable identified rural poor families to cross the poverty line by providing productive assets and inputs to the target groups. The assets which could be in primary, secondary or tertiary sector are provided through financial assistance in the form of subsidy by the government and term credit advanced by financial institutions. The program is implemented in all the blocks in the country as a centrally sponsored scheme funded on 50:50 basis by the Centre and State. The Scheme is merged with another Scheme named *swarnjayanti gram swarozgar yojana* (SGSY) since 01.04.1999.

Chapter 4

Communication

1. Communication is a process by which message are transferred from source to receiver – rogers and shoemaker
2. Communication is process, as the process of sending message and receiving message through channel which establish a common meaning between source and receiver – van Dan ban and Hawkins
3. Communication is a conscious attempt and a social process
4. Four basic functions of communication: Informative function
    a. Command or instructive function
    b. Influence or persuasive function
    c. Integrative function
5. Information is the basic requirement of adapting oneself to the environment
6. Code is a system of signals for the communication
7. Schramm model is particularly relevant is mass media
8. Task of communication is to provide powerful incentive to change
9. Extension agent= communicator
10. Universities = originator of the message or source
11. To be a good communicator the thumb rule is GO TO PEOPLE AND LISTEN TO THEM
12. A good message should clearly state that what to do, how to do, when to do and what would be the result

13. To produce desirable change is human behavior the message must be motivating
14. Channel of communication constitute the medium through which information flow from a sender to one or more receiver
15. Channel of communication can be classified into several ways:

    According to form:

    a. Spoken – farm and home visit, farmers call, radio talk *etc.*
    b. Written – personal letter, farm publication, newspaper *etc.*

    According to the nature of personal involved

    a. Personal localite: important in traditional social system, face to face and belong to same environment
    b. Personal cosmopolite: face to face, not belong to own environment. *E.g.* extension agent
    c. Impersonal cosmopolite: neither faces to face nor belong to same environment. *E.g.* mass media

    According to the nature of contact with the people:

    a. Individual contact
    b. Group contact
    c. Mass contact
16. When too many channel enter in series then noise enters
17. So if there are many channel present they must be presented in parallel
18. Treatment means the way message is processed so that information gets across the audience
19. Treatment and presentation of a message shall depend to a large extent on the choice of channel and the nature of audience
20. Audience appeal is a psychological bridge to getting a message delivered
21. If communication is to be successful it should be target oriented
22. Audience is not a passive recipient of message, the individual are rather selective in receiving and processing and interpreting message
23. Selective exposure – klapper 1960
24. Audience segmentation – Rogers (heterophilous population into homophilous one)
25. Audience response is the ultimate objective of any communication function
26. Understanding is achieved when can attach meaning to the facts, see the relationship of facts to each other and to the problem

27. People not act on the fact/knowledge alone they act when they gain understanding
28. Feedback is source oriented and exerts control over future message
29. Feedback affects communication fidelity
30. If there is a favorable response of a message by an audience is received then extension agent should supply critical inputs and the credit since these are very important at these stage
31. Feedback act as pathfinder for the need based research
32. Consumer behavior exercise a major influence on communication strategies for marketing
33. Act of communication has at least 3 phases;
    a. Expression
    b. Interpretation
    c. Response
34. Individual frame of reference – person's behavior in a particular situation.
35. Perception- process by which an individual receives a stimuli through various senses and interprets it
36. Noise is inversely proportional to the fidelity
37. Communication fidelity can be explained as extent of desirable changes in receivers behavior as a result of communication

    Desirable changes are in the receiver's knowledge, attitude and Action
38. Use of more than one channel of communication parallel will help the communication even more
39. Empathy is indispensable skill of people moving out of traditional setting
40. Heterophilous interpersonal links often connects two cliques
41. Heterophilous interpersonal links are often useful in carrying information about the innovation
42. Homophily accelerates diffusion, but it confers it into a limited numbers
43. Publicity = truth + propaganda
44. Development is a widely participatory process of directed social change in a society intend to bring about social change and material advancement of majority of the people through their gaining greater control over environment
45. Development communication is the use of communication to further the development
46. Opinion leaders are in close conformity with the social system

47. Thayer pointed out at least four level of analysis of human communication
    a. Intrapersonal
    b. Interpersonal
    c. Inter-organizational
    d. Intra-organizational
48. Message is the actual physical product of the source encounter
    - ☆ All communication takes place around the message
    - ☆ Part of persons behavior which perceived by another as having implication or meaning for himself
    - ☆ If the content of the message is changed the phenomenon of the communication will not change
49. Extension education in which means is as important as the ends
50. All other ingredients in a communication event purposely revolved around the receiver
51. Process is a phenomenon which shows a continuous change with time, or any continuous operation or treatment
52. Communication is an essential continuum to human beings
53. Berlo views communication is a dynamic, ongoing, ever-changing continuous act which have no beginning nor any end and can't be freezed at any time
54. Communication is a byproduct of any interacting individuals
55. Communication takes by many means at many places for many reasons and with many people in many ways
56. Communication is process both reciprocal and alternating
57. At organizational level new factors emerge these are
    a. Law of large nos.
    b. Implication on the behalf of
58. Rogers said that an organizational structure can restrict the accessibility off receivers to sources and helps to decrease information overlord
59. Communication takes place a. within participant b. in between them
60. Within communication determined by:
    - ☆ Individuals take into abilities
    - ☆ Take into susceptibilities
61. Communication is the process of organizing and converting raw sensory data into functionally consumable or process able units of information

62. Information is the ultimate element of the human communication
63. 17th century faculty psychology – distinction between mind and soul
    - ☆ One purpose is intellectual (informative)
    - ☆ Another purpose is emotional (persuasive)
    - ☆ 3rd purpose is entertainment
64. Schramm – immediate and delayed reward
65. Festinger – purpose of communication is consumatory and instrumental
66. Berlo – all communication behavior has its purpose and its goal as the production of response

    Purpose of communication should be behavior centered
67. Aims and function of mass communication as surveillance interpretation and prescription, transmission of culture and entertainment
68. The question is not whether the communication will take place but with whom and with what consequences
69. Take into account abilities are a function of the capacity of an individual conceptual system to ascribe meaning or significance to certain pattern and sequence going in his internal and external environments
70. Living is largely a matter of communication – smith
71. Intercommunication might thus be viewed as a process of mutual regulation and control
72. The immediate and ultimate consequences in intercommunication are determined by process which occur within each participants not by what goes between them
73. Communication needs and function of organization are fulfilled by at least 3 system
    - ☆ OPERATIONAL COMMUNICATION SYSTEM - through this data about task related activities in communicated within organization from its different sources generating it
    - ☆ REGULATORY COMMUNICATION SYSTEM - referred as internal or external dynamics.
    - ☆ Messages are thus the natural consequences of the problems definition behavior of the organizational managers
    - ☆ MAINTAINENCE OF DEVELOPMENT OF COMMUNICATION
74. INTERORGANIZATIOONAL COMMMUNICATION IS – inter-organizational data transport linkage
75. Communication effectiveness- means a communication encounter that has produced desired result

76. Measures of communication effectiveness:
    - ☆ Effectiveness of communication encounter (achieving commonality between communicator and receiver)
    - ☆ Effectiveness of communication performance (the actual effect on other person after a longer conversation)
77. Thayer (1968) – originators interpretation of the message may not be the accurate than the receiver In fact originate has more reason to distort the message
78. Desired conduct on the part of receiver has been operationalized by different factors Berlo identified 4 factors
    a. Attitude
    b. Knowledge
    c. Communication skill
    d. Position in a social system
    e. The four factors also effects the effectiveness of the communication
79. Thayer also mentioned factors which effects the communication
    a. Communicator receiver relationship
    b. Purpose or intentions
    c. Roles and rules
    d. Organization on the behalf which they engage themselves
    e. Also value such as intellectual, attitudinal, communicational *etc.* brought by each participant
    f. And commitment to cooperate within communication act
80. Jha operationalized effectiveness by:
    a. Attitude changed
    b. Knowledge gained
    c. Knowledge retained
81. Thayer described communication on the context of message treatment

    So according to him communication effectiveness is a function of
    a. Comprehension
    b. Validity
        1. Consonance (legit)
        2. Credibility
        3. Conguigrity
    c. Utility

82. Deustch definition of model is clearly based on nature and process
83. In social science models are of two types
    a. Structure oriented
    b. Process oriented
84. Broadbeck and bross: models have a ambiguous standing in the philosophy of science
85. Creation of significant things to say is termed as a rhetoric invention
86. Five process involved in the study of communication (Aristotle)
    a. Invention – message preparation
    b. Organization – arrangement of some persuasive manner to fit the audience
    c. Language and style
    d. Memory
    e. Delivery
87. Discovery of the available means of persuasion as his definition of the whole art of persuasive communication
88. Rhetoric theorist were interested in ability to effectively communicate
89. Laswelll (1948) was one of the earliest social scientist to come up with rudimentary model of communication.

    Laswell model is – political communication, political propaganda and political symbolism.
90. Most basic model – stimulus response model (interpersonal communication)

    According to S-R theory communication may not be described in linear terms it is actually a complex system.
91. Schramm's model –
    a. Sender and receiver must be tuned together
    b. Each person in the communicating process is both encoder and decoder
    c. This model emphasizes on the organic nature of communication
    d. It is misleading to think that communication is starting somewhere and ending somewhere, rather it is a continuous process
92. Shannon Weaver's Model-
    a. Information theory of communication
    b. 3 level of problem in communication

1. Technical
2. Semantic
3. Influential

c. The mathematical theory book provide a visual model of Shannon weaver communication system in relation to the "electronic media"

d. Noise is introduced for the 1st time

e. Information is basically what could be said not what is said.

f. Information is defined as logarithm of the number of choice

g. Relative entropy = actual entropy/maximum entropy

h. English language is 50 per cent redundant

i. Channel capacity – amount of information that can be transmitted per unit time

j. To transmit a given quantity of information a definite product of band width and time

k. Maximum error free capacity of noise free channel

C= W log2 [P+N/N] bits per second

P= average signal power

N= noise

W = band width

l. Efficiency can be increased by
   1. Increasing band width
   2. Increasing signal power

m. Efficient coding is that which matches the statistical characteristics of the information source and the channel

n. Coupling – any system that couples two other system is called the gate keeper

o. Shannon weaver model doesn't deal with a single message, but with the statistical characteristics of the whole message assembly

93. Berlo – one of the most widely used model; based on behavioral theory
    a. Communication is a continuous process
    b. Channel include – seeing, hearing, touching, smelling, tasting
    c. Berlo's model is a linear model which emphasizes on communication
94. Leagans emphasized on the treatment of message and audience response
95. Hoveland persuasion model is receiver-listener oriented

In these both the communicator and the message are seen as observable communication stimuli

Hoveland was primarily interest in predicting attitudinal change

Attention, comprehension and acceptance are the mediating process the hoveland model

96. Westley and Mclean model (Mass media), earlier developed by Newcomb
97. Riley and Riley model: mass communication is one social system among many society

    It is also known as social system model

    Model according to authors, seeks to fit together the many message and manifold individual reaction to them within a social structure and process
98. Gerbner's model: also known as perceptual model
    a. Primarily concerned with perception and production of messages
    b. Someone perceives an event and reacts in a situation through some means, to make available materials in some form and context conveying content with consequences
    c. McQual and windhall – stated that model is subjective, selective, variable and unpredictable
    d. Numan communication are open system
99. Subject matter of linguistics is spoken language Therefore it deals with description of language system Orientation of linguistics theory tend to be social Meanings are in the people, not in words When we emphasize on conative meanings, fidelity decrease but readability increases Linguist: help our thinking about codes Two dimensions of linguistic theory
    - ☆ Psycholinguistic theory: behavior of an individual using of language It also provided us with the nature of coding behavior
    - ☆ Sociolinguistic theory: relationship between variation in language usage and the personal and contextual circumstances of usage come under the realm of sociolinguist

    Sociolinguist pays relatively more importance to communication competence
100. Theory of mass communication: defleur (1960)
    a. Theory of individual differences
    b. Psychodynamic model of communication
101. Medium is the message – McLuhan
102. Content of the message are relatively less important than the way they are received

103. Technological development are extension of man
104. Printed media encourage a man to think in a linear, whereas electronic media in a mosaic way
105. Hot media – Radio, Movies Cold media – Telephone, TV
106. Model of opinion change – Kelman
    a. 1st phase: linear, one way, manipulative
    b. 2nd phase: organic, circular, two way
    c. 3rd phase: interactional, in a specific social cultural context
107. Communicator is the first person to give expression to a message
108. Communication behavior – communicator expression of results from information seeking, information processing and information disseminating behavior
109. Osgood et.al. emphasizes on importance of silence in the communication and Osgood also discovered the brainstorming
110. Berlo – five verbal communication skills

    Broadly divided into 2 types

    a. Encoding: writing, speaking, painting, drawing, gesturing
    b. Decoding: reading, listening
111. Communication skill determined communication behavior
112. Attitude affects in the way it communicates
    - ☆ Attitude towards self
    - ☆ Attitude towards subject matter
    - ☆ Attitude towards receiver – Aristotle called this perceived characteristics of the speaker ethos, quality in the speaker that is more appealing to the audience
113. Status and prestige are the common relationship factors, relationship are the basic pre requisite of any inter communication
114. Communication always occurs in the service of some need of psychological system
115. Rules are psychological, sociological and anthropological uses

    Role is a type of rule
116. COMMUNICATION COMPETENCE:

    Six distinguishable characters to classify competence

    a. Fundamental competence
    b. Social competence

c. Interpersonal competence
d. Linguistic competence
e. Communication competence
f. Relationship competence – examines the relationship between behavior ad outcomes such as perception of appropriateness and effectiveness or social skillfulness

First three skills are concerned with achievement of outcomes; fourth and fifth skills are message focused rather than outcome focused and attempt to explain the message behavior. Relationship competence more concerned with communication process and functional outcomes.

117. Communication competence is broader in scope than linguistic competence
   - ☆ Ability to perform as well as the knowledge of how to perform
   - ☆ Ability to process information cognitively and ability to explain and predict human behavior
118. 2 dimension of competency
   a. Knowledge
   b. Skills
119. Communication competency has 5 major components (spitzberg and caupauch)
   a. Knowledge
   b. Motivation
   c. Skills
   d. Context
   e. Outcomes
120. Competence is both molar and molecular
   a. Molecular: provide specific communicative indicators of competence as well as reference point for skill enhancement
   b. Molar: provide evaluative outcome indicators
121. Sequence of extension communicator roles – 7 (Rogers and shoemakers)
   a. Develop a need for change
   b. Establishes a change relationship
   c. In diagnosing the problem
   d. Create an intent for change
   e. Translate the intent into action

f. Stabilizes change and prevent discontinuance

g. Achieving a terminal relationship

122. Communication activities (bhanja) – as a percentage of utilization of extension agent

a. Writing and reading routine official communication (21.17 per cent)

b. Interpersonal communication with staff and visitors (16.11 per cent)

c. Reading writing other than routine work (13.57 per cent)

d. Formal and informal meetings (12.23 per cent)

e. Restful thinking (intrapersonal communication) (11.46 per cent)

f. Giving dictation (11.14 per cent)

g. Attending incoming and outgoing calls (5.57 per cent)

h. Others (9.16 per cent)

123. Communication IQ – refers to an extension executive's understanding of the intricacy of communication effectiveness

a. Agroexecutive's score: 5.39 (equal to civil servant and forest officials)

b. Bank and commercial organization: 5.0

124. 87.5 per cent of the agricultural extension administrators have a medium organizational leadership.

125. Thayer – communication economy concept

126. Communication effectiveness depend upon: 4 characteristics of communication

a. Communication behavior

b. Role perception and role performance

c. Relationship with audience

d. Program content

127. Effective oral communication is a function of clear objectives, good functional relationship and effective delivery

128. Effective talk rate - 100 words/minute (140 for radio and 160 for lecture)

129. Effective communication in extension is assumed to be matter of promoting behavior change

130. Code: is anything which has a group of elements (vocabulary) and a set of procedure for combining those elements meaningfully (a structure)

131. Content: content as the material in the message selected by the source to express his purpose

132. Treatment: arranging both code and content in order to achieve comprehension on the part of receivers
133. Message is not the subject matter, rather its generalized idea what subject matter means
134. Treatment deals with designing of methods for presenting message
135. Strategies for treatment of message means techniques employed within the situation provided by the message and channel
136. Purpose of treatment: To make message interesting, clear, understanding and realistic to the audience
137. Extensity: the larger the stimulus more likely it will be noticed
138. Avoidance of condemning is one of the most basic method of speaking
139. Two variables that effects strategies for effective treatment of message:
    a. First is the amount of credibility the communicator enjoys in the mind of receiver
    b. Receiver initial attitude towards message
140. Burgoon and Burgoon (1975) – general rules for preparing a persuasive message.

    If audience is initially in agreement with message then a one sided message may suffice
141. People respond to a message that are
    - Reliable
    - Realistic
    - Relevant
    - Understandable
142. Message efficiency - The relative utility of the message to the receiver together with the relative ease or difficulty of the receivers comprehension of the message may be considered
    - Message efficiency is encounter specific depend on
        a. Participants
        b. Situation
        c. Timing
143. Hypodermic needle model is based on S-R model
144. Main disadvantage of noise is wastage of time and energy
145. Loss of message is maximum in – mirage distortion
146. Channels – physical bridges between the sender and the receiver of messages

147. Media means of transporting data in a particular way, what Mcluhan calls extension of man's abilities to acquire and disseminate consumable data

148. 3 dimension of concept of channel namely:

    a. Modes of encoding and decoding messages such as our speaking and hearing mechanism

    b. Messages vehicle such as sound waves

    c. Vehicle carriers that support the sound waves during transmission

149. Communication context in interpersonal channel is face to face whereas in mass media is interposed

150. Possible effect of interpersonal channel – attitude formation

151. Possible effect of mass media channel – knowledge change

152. Efficiency of communication channels predicts to their use and creation in the emergence of communication system between and among people

153. Novelty effect is one of the main reason for the success of media forum

154. The response of receiver to a message seems to be a function of the whole personality

155. Berlo (1960) – pointed out that response of a receiver depend on

    a. Communication skills

    b. Attitudes

    c. Knowledge level

    d. Position in a social system

156. Receiver is the most important link in the communication process

157. Communication behavior- refers to the degree to which a farmers expose himself to various information sources communicating agro technology the extent to which he processes the information and the extent to which he passes the information to others Sandhu and lal surveyed that 36 per cent of the farmers have a low communication behavior 44.3 per cent of the farmers have a medium communication behavior 20.67 per cent of the farmers have a high communication behavior Source utilization pattern of farmers was determined in their importance by their

    a. Social participation

    b. Education

    c. Socio economic status

    d. Value orientation

158. Models of social interaction:

    a. Least satisfying – wheel

b. Moderately satisfying – chain (interaction is possible but it is prone to distortion)

c. Greater amount of satisfaction – Star (quite efficient for cognitive tasks but least efficient for production tasks)

** do read all the models and remember all the diagrams

The fewer the networks greater will be the fidelity of information

In an organization, the more restricted the communication channel the more efficient is the flow of communication

159. Communicator: receiver relationship can be measured in terms of

    a. Physical interdependence
    b. Action reaction interdependence
    c. Empathy
    d. Interaction
    e. Heterophily

160. Feedback makes both the communicator and receivers truly interacting members of social system

161. When a source receive feedback that is rewarding, he continues to produce same kind of message

162. Sender experience contribution more than receiver experience in improved accuracy of communication

163. Feedback is an aid to accuracy in an interpersonal situation

164. Empathy is a process which refers to the ability to project ourselves into people's personalities

165. Role taking theory suggests that the concept of self is learnt through communication

166. Empathy accuracy - Group α 1/accuracy, or group size is inversely proportional to accuracy of communication.

    ☆ Prior communication α accuracy
    ☆ Sensitivity α accuracy
    ☆ Motivation α accuracy

167. Interaction – mutual and reciprocal influencing of each other's behavior. Also it is the process of reciprocal role taking

168. To the extent we are in interactional situation our effectiveness, our ability to effect and to be effected by others increases

169. Interaction is the ideal communication and also the goal of human communication

170. Heterophily marked as the basis of relationship between communicator and receiver in an extension situation

171. In order to achieve a communication situation in which there is a full and effective interaction of hearts, body, and mind

172. 4 approaches of mapping organization's public

    a. Social system analysis - loomis (1960)

    b. Social stratification - williams (1969)

    c. Social differentiation - bertrand (1967)

    d. Cultural analysis - spicer (1952)

173. Wright (1966)

    a. Mass communication are rapid and consumed immediately

    b. Output of the mass media are considered to be expandable

    c. Mass communication is – public, rapid and transient

174. In mass communication, from the POV of the communicator, the message is addressed " to whom it may concern "

175. 1965 – ten intensive farm and home units are established at selected AIR stations

    To feed factual and technical information to the farming community

    At present broadcasting network covers 93 per cent of the population spread over 82 per cent area.

176. Radio wave have a sense of immediacy

    a. Through on spot broadcasts or through simulated broadcast radio can overcome the barrier of space and time

    b. Not a very suitable medium for giving precise and detailed information, but a very suitable in achieving a favorable attitude and motivating learners to action

177. Television also represents selective attention

    TV is an ad lib medium

178. Videotex – information system with which text and graphics information is transmitted and then received either by videotext TV or an ordinary TV with adapter Videotex coupled with microcomputer permits editing as well as retrieval of information on a videotext page

179. Teletex – videotext transmission broadcast signals is known as teletex

    a. Teletex is a system that links computer to a TV

    b. One way system

c. Indian TV in teletex – 1985, known as intex service

d. To telecast several information on stocks, weather *etc.*

180. Interactive videotex: videotex transmitted by telephone cables, data lines on a private network

 a. Fully interactive and can even print a hard copy of any page that it appears

 b. Two way nature

181. First experiment with satellite TV broadcasting was conducted in India in 1975-76, when SITE had come

 a. Site launched in 34000 villages

 b. Spread over 6 states

182. INSAT 1A – 1982

183. INSAT 1D – 1990

184. INSAT 1B – technical life of 7 years

185. CAI is a natural outgrowth of principle of programmed learning and can be used to educate 40,000 learners

186. Computed stores massive amount of ORGANIZED information

187. Rogers and Kincaid defined KEY COMMUNICATORS

188. Sociometric method needs a large no. of respondents and a complex statistical analysis and also it has a higher validity

189. When norms of a system are more modern key communicators are monomorphic] Monomorphic is the tendency of a communicator to act on a specific topic

190. Shankariah's findings on Indian farmers are –

 a. When is village society is more modern then we should choose best farmer as the key communicator

 b. When village society is generally traditional then we should choose Social stars as key communicator

191. According to Sandhu – key communicator age is around 30-50 years

192. Characters affecting key communicators role –

 a. Social participation

 b. Education

 c. Value orientation

 d. Socioeconomic status

193. Adequacy of language to serves as a medium of communication stem from the primitive emphatic response of an organism
194. Radio is an extended human voice
195. Radio broadcasting history:
    a. 1924 – 1st broadcast madras presidency club
    b. 1930 – formation of Indian broadcasting service
    c. 1936 – the name changed from IBS to AIR
    d. 1957 – then name again changed from AIR to Aakashvani
196. Growth of mass communication
197. Mass media tend to produce product uniformity
198. Mass media convey legitimacy of status
199. Communication of meaning is a complex problem
200. Mental functioning of recognition, verbatim recall and suggestibility seem more effectively aroused in listening
201. Critical attitudes and discriminative comprehension are formed by Reading
202. Uncommitted people become propaganda target
203. Receiver responds to the status conferred by mass communication
204. Basic long term response of communication is slow infiltration and coloring of individual responses
205. Mass communication increases commonality if experience among public
206. Effects of books, newspaper, magazine *etc.* have effects which can be classified into:
    a. Instrumental
    b. Prestige
    c. Reinforcement
    d. Enrichment aesthetic experience
    e. Respite
207. Mass communication intensifies propaganda conflicts.
208. Level of communication: four level
    a. Conventional level: communication of convention level is fairly impersonal. It just open the door of further communication

       It is just merely an attempt to acknowledge the presence of another person

       e.g. how do you do?, good morning *etc.*

b. Exploratory level: also fairly impersonal, but it become fact and problem oriented

   After initial communication people start discussing things, people, weather *etc.*

c. Participative level: at the participation level, people talk about themselves and engage in self-disclosure. They express their feeling, describe their own ideas. Though the communication in this stage shifts to personal level, it is still expressed in a very safe way

d. Confrontational level: this type of communication is common among friends lovers and married people rarely found in organization

   In this level people reveal themselves intimately to one another in ways that involve risk. This communication evolved from deeply felt mutual understanding it is found in mutual relationship

COMMUNICATION METHODS

1. Method – procedure or process for reaching an objective Communication method mostly depend – number and location of audience and the time available in communication
2. Individual methods benefits:
   a. Enhances effectiveness of group and mass media
   b. Helps in selecting demonstrations and local leaders
3. Group methods: group participation and formation of group opinion is important
   a. Small group – 15 -25 persons
   b. Medium group – 25 -50 persons
   c. Large group – 50 -100 persons
4. Group method benefits (Odell junior)
   a. Efficiency
   b. Effectiveness
   c. Equity
5. Mass method – normal group boundaries get obliterated
6. Size of audience
   a. Mass meetings – few hundred
   b. Campaign and exhibitions – few thousands
   c. Newspapers, radio, TV – millions

   Mass media helps to reinforce previous learning

7. Types of ecosystem
    a. Friendly: favorable with little or no risk
    b. Inhospitable: unfavorable with an appreciable risk
    c. Hostile: hazardous with high level of risk and uncertainty
8. In interpersonal communication message are relatively high socio economic content (singhal and rogers)
9. Basic objective of extension teaching is to create opportunities for effective learning in order to secure changes in the minds of the learners
10. Ensminger (1957): before an extension worker can become efficiency in the use of methods he must know what methods are available when to use a given method and become effective in using methods
11. Wilson and Gallup divided extension teaching methods
    a. According to their use
        1. Individual
        2. Group
        3. Mass
    b. According to their form
        1. Spoken
        2. Written
        3. Visual
        4. Audio visual
12. Farm and home visit:
    a. To sustain interest
    b. To gain confidence of farmer and homemaker
    c. Helpful in giving urgent info
    d. To discuss individual or village problems
    e. To reinforce other methods
    f. First part of meeting should discuss some possible which farm can take to improve their condition
    g. Second part of the meeting is to encourage action and offer him help in becoming competent to take action
    h. Very helpful and useful in helping farmers get through the evaluation stage of adoption and in teaching skills
    i. Good source of new stories

13. Farmer's Call:
    a. Three hours and twice a week would suffice the purpose
    b. It enhances prestige and enhances professional satisfaction of the extension worker
    c. Provide an insight into current farmers problems
14. Personal Letter: should not consider as a substitute for personal contact.
    a. In personal letter extension agent writes letter to farmer
    b. Personal letter writing formula – McCormick
        1. Shortness
        2. Simplicity
        3. Strength – use specific concrete words and active words
        4. Sincerity – personal words and sentences
    c. Information sent through mail request should be answered promptly and as affectively as possible
    d. Information send through personal letter directly related to the problem
15. Agriculture clinics: - Team of SMS, very expensive method
    a. Two hours set aside for the concerned specialist to visit the site for on farm diagnosis and guidance farmer whenever necessary
    b. Method can be used where agricultural universities farm advisory services are operative or places where SMS are provided via T&V system
    c. Helpful in specialized guidance and provide farmer specific technology transfer
16. Minikit trail: - method of determining the suitability or otherwise of a new practice in farmers situation
    a. On farm participatory technological development process in which farmer choice and farmer opinion about the practice is the most important
    b. Selection of innovative farmers are one of the need for conducting trials
    c. First stage a new and improved practice passes through before it is taken up for results demonstrations or methods or recommended for a large scale adoption
17. Result Demonstration to be effective should be integrated with total extension program
    a. Build confidence in farmers and extension agent

   b. Helps to develop innovation leadership
   c. Selection of responsible and cooperating farmers are important
   d. Prepare a calendar of operation
   e. Conduct field day or farmers day around all successful demonstration
   f. In this more time energy and funds are requested for extension work
   g. Basic idea of using demonstration
   h. Gabriel de tarde theory of instinct imitation
   i. New cultural traits originate from one creative individual and imitated by the others
      1. Short duration RD – weedicide, pesticide
      2. Long duration RD – new seed variety

18. Peripatic team visits:
   a. Team of 3-5 SMS led by extension officer
   b. Visits a village on a pre-drawn schedule
   c. Here team of specialists travels from one village to another
   d. Presentation made by SMS is very brief more emphasis is placed answering question and organizing debates, later individual guidance is also provided by examining problems samples or by visiting problem field or sites *etc.* (that's how it differs from group method)

19. Method demonstration: it is essentially a skill training program.
   a. Oldest from of teaching
   b. Where emphasis on effectively carrying out a job, which shall clearly indicate upon the result.
   c. Steps in method demonstrations:
      1. Gather information make a desired plan
      2. Select demonstrations
      3. Carry out demonstrations
      4. Follow up

20. Group meetings:
   a. Aim at collective decision making and improving individual decision making
   b. Aims at people participation and facilitates program implementation
   c. Develops the ability to face adverse situation
   d. Convenient size: 15-25 which may be extend up to 50

e. Help in formation of opinion about a specific topic
f. Act a valve to reduce tension
g. Extension agent shall act as motivator and resource person
h. Mosher- meeting opportunities can be lost if individual launched lecture
i. Active people find it hard to sit more than one and half hours

21. Small group training: it is a technique of imparting specific skills to a group of people who need them by creating an appropriate learning situation.
    a. It is actually for small group of people
    b. Motivate people to adopt new skills
22. Field day or farmer's day: method of motivating the people to a new practice by showing what actually can be achieved by applying the practice under field conditions.
    a. Can be held in farmers field or research station
    b. If group is large it can be divided into small groups of 20-25 each
    c. Motive farmers to adopt the practice by showing its performance and profitability under field conditions
    d. Arrange display of exhibits diagrams chart *etc.* near the place of meeting
    e. Invite few visitors to give their reactions and their answer the question raised
    f. Major limitation is that it doesn't allow in depth learning and field days cannot be hold frequently
23. Study tour:
    a. A group of 30-50 persons may be convenient for a study tour
    b. Maximum no. is 80-100
    c. This method give farmers an opportunity to observe a practice already being adopted in an actual situation
24. Under T&V system there is a provision to arrange media forum
25. Scientific attitude and judgment making can be built by agricultural games
26. For using these games 2 points should be kept in view – Swaminathan
    a. Adopt both form and content of the game to local situation
    b. Making, designing and preparation of such games is itself a worthwhile educational activity

27. Snake and ladder – Mishra (1974)
    a. Up to 4 persons can play this game
    b. Ladder indicates – desirable practices to be followed
    c. Snake indicates – extent of damage caused by a particular practice
28. Road to agricultural prosperity:
    a. Developed by university of Udaipur
    b. Four person can play at once
    c. Steps illustrated important aspect of farming
    d. Played by gotties and dice
29. Playing cards – directorate of extension, ministry of agriculture, India
30. Test yourself: following the principle of braino games

    Prepared by directorate of extension MOA, GOI
31. Mass media bale to create a psychological effect that people will not like to be left behind from the other members of social system
32. Campaign: intensive teaching activity
    a. Undertaken at a opportunity time for a brief period
    b. Focusing attention in a concentrated manner on a particular problem so as to stimulate the widest possible interest in the community
    c. Campaign must be directed towards the solution of a problem which the people must recognize
    d. Emphasize only one idea at a time
    e. Method is used to create conducive psychological climate for the adoption of an improved practice in quickest possible time
    f. Used to stimulate community as a whole
    g. Series of method demonstrations can be done to show people how to adopt a practice
    h. Credit people rather than crediting extension agency
    i. Evaluation to be done based on the objective of rather than activities undertaken
    j. Single day campaign – water for life
    k. Few weeks campaign – rat control or family planning
    l. Few months – vanmahotsav
33. Exhibition: systematic display of models, specimens, charts, photographs, pictures, posters, information in a sequence around a theme to create awareness and interest in a community

a. Suitable for reaching all type of people
b. Promotes visual literacy
c. Acquaint people with better standards
d. Create interest in wide range of people
e. Motivate people to adopt better practice
f. Forming a steering committee and a suitable subcommittee with the specialist
g. Display exhibits about 50-60 cm height above the floor, upto a height of 2 meters
h. Purpose is to catch attention, convey an idea and stimulate action
i. Good exhibitions build around a single theme with a minimum supportive information
j. Lettering to be simple, clean and bold
k. Use 2 to 3 bold colors against a neutral background
l. Place the center of interest near the eye level – five feet above the ground
m. Information below 3 feet and above 7 feet can't be seen properly

34. Farmer fairs: fairs should be held for one to two days, twice a year corresponds to the period when summer and winter crops are ready to harvest within a month
    1. Agro home industrial exhibitions: commercial firm are also invited to their stalls and demonstration, to provide information to the visiting farmers
35. Farm publication:
    a. Leaflet: single sheet of paper, preliminary information, generally free of cost
    b. Folder: single printed sheet of paper of big size, folded once or twice, essential information is given, free of cost
    c. Bulletin: bounded booklet with a no. of pages, contains comprehensive information, small price may be fixed
    d. Newsletter: miniature newspaper in a good quality paper, information regarding activities and achievement of the organization is given, fixed periodicity and free of cost
    e. Journal, magazine – information related to various topics of interest not only for farmers but for extension officers also, fixed periodicity, pre- payment of subscription

Devote the first paragraph to highlight economic and other benefits. This may be printed in bold letters

f. Circular letter: used as a follow up of several others extension agent

g. Good circular letter have three essential parts

    1. The introduction for rapport building and linking new message with previous one
    2. The information of the message
    3. The appeal to the audience

h. Newspaper: newspaper are printed on a special type of paper known as newspaper

    1. Small newspaper are also known as rural newspaper

36. Behrens and Evans (1984) – 6 ingredients which determine worthiness of an information

    a. Timeliness
    b. Nearness
    c. Consequence
    d. Human interest
    e. Prominence
    f. Editorial policy

37. Agrinews in any newspaper is around 3.5 per cent, in the Hindu newspaper agro news is on Thursday.
38. Radio doesn't hamper daily work
39. Objectives of farm and home broadcast

    a. Broadcast technical information of a continual basis in alignment with the package of practice in respect of important crops of a particular area along with information about services
    b. To inform and educate the rural women on their efficient partnership to change from traditional to modern farming system

40. Involvement of extension agent with the radio may of of two types:

    a. When a program is made or field recording
    b. When extension agent is invited to deliver talk

41. Duration of broadcast is about 10-20 minutes
42. Delivering a talk:

    a. No. of participants – 3-5
    b. Duration of straight talk – 5-10 minutes

   c. Duration of discussion – 10-20 minutes (duration of broadcasting has same timing)

43. Journal article has headlines, leading sentences or paragraph and a body of contents. We should use action verbs, simple words.

    Body of content is the most important part

44. Radio rural forum: forum member listened bi weekly broadcasts on all India radio for a duration of about 30-45 minutes/day

45. Forum has a chairperson, a convener, and 10-20 members

46. The format of farm teleprogramming may be interview based, discussion based of the documentary type or recording or real event type or performance *etc.*

47. Telescript has a video column on the left and audio column on the right

48. TV is an ad lib medium and a close up medium

49. Non projected visual aid:

    a. Presented:

       1. Flash card
       2. Flannel graph
       3. Flip chart
       4. Maps

    b. Displayed:

       1. Photograph
       2. Posters
       3. Wall chart
       4. Bulletin
       5. Exhibits

    c. Literature:

       1. Illustration
       2. Leaflets
       3. Comics

50. Dale classified audio visual aid according to their learning contribution

    >>>From maximum to minimum<<<

    a. Direct purposeful experience
    b. Contrived experience
    c. Dramatized experience

d. Demonstration
e. Field trips
f. Television
g. Motion posters
h. Recording, radio, still pictures
i. Visual symbols
j. Verbal symbols

51. Drawing may take a form of – illustrations, graphs *etc.*
52. Words are numbers used in titles and explanation
53. Pictures are used in – sketches, symbols, photograph
54. Principal graphic: drawing
55. To keep the cost down no. of color should not exceed – two to five
56. Layout: basic composition or arrangement of element or constituent parts embraces the arrangement of graphic aid, illustration, script, color and white spaces
57. Two main qualities of good layout: balance and attractiveness
58. Formal balance – symmetrical arrangement of element of display
59. Informal balance – objects or elements appeared to be balanced, placing the main idea near the optical center
60. Spacing should be optical
    a. 100 ft. → 3.5 inches
    b. 80 ft. → 2.7 Inches
    c. 60 ft.→ 2.1 inches
    d. 40 ft.→ 1.4 inches
    e. 20 ft.→ 0.7 inches
    f. 10 ft.→ 0.5 inches
61. Demonstration poster lettering – 1.5 inch height
62. Since most of the audience lies around – 48 -50 ft. away
63. Titles – 1.5 inch
64. Text – 1.0 inch
65. Flash card – simplest of all graphic cards
    a. 10-25 people – 10" – 12"
    b. 30-50 people – 15" – 20"

Total of 10-12 cards are normally flashed, should not be used in large gathering

Captions in flash card should be brief – it is better to limit them to 5-6 words

66. Flannel graph: the color of basic flannel board is – black, gray, dark green and dark blue
    a. Visuals are particularly effective in presenting detailed concept
    b. Action like effect in placing flannel on the board
    c. Take a piece of plywood of about 24 by 30 inch
67. Chart: chart are a combination of such material, which together are more likely to represent clear visual summarizes of important facts, concepts and relationships
    a. Types of chart:
        1. Tree – branches represents developments and relationship a reverse is useful in showing how a great variety of elements combined to form one important product
        2. Flow chart – useful when purpose is to show cycle of plants growth
        3. Outline chart – organization of content into key point and supports which communication may do on the chalkboard
        4. Tabular chart – a unique value of this chart is the ability to show time relationship
    b. Generally 20 inch by 30 inch are good for normal purpose
    c. Lettering: Titles – 2.5 inch, subtitles– 1.5 to 2.0 inch, letter body – 1 to 2 inch
    d. Chart can be used as posters by cutting the picture from the chart and adding a suitable caption. Can be used in exhibitions
68. Graph: defined as a visual representation of numerical data Shows relationship in the data such as trends and variation from the normal
    a. Line graph – precise and accurate of all graph, useful for showing trends and relationship between two sets of data. It should be only when considered amount of data are plotted or only when data comprises of continuous series, which overtime clearly shows progress
    b. Bar graph - simplest of all the graph (best for farmers)
    c. Pie graph
    d. Pictorial graph – pictorial statistics are the 1st popularized by Ouo Neurath

e. Area and solid figure graph- no. of units should not be more than 4-5
f. Circle graph (pie graph) – read most accurately when used to compare parts of a whole
g. Multiple area column graph – read least accurately
   ✰ Bar graph along followed by line graph are the best for use with farmers (powers)
   ✰ National project in agriculture communication at Michigan state university concluded that
   a. Accuracy of estimate drops when number of dimension of the graph increase (2D to 3D)
   b. Accuracy drops when number of element in a graph increases

69. Posters: modern posters were born in 1867 in Paris by jules cheret
    a. Color provided – force, contrast, and attractive
    b. Good posters require a center of interest
    c. Posters strategically placed and frequently changed are highly effectively
    d. Posters may be continued to be valuable even after their initial impact has passed. This may be called as ' atmosphere or environment use'
    e. Size: standard size of the posters
       1. 10*15 inches
       2. 15*20 inches
       3. 18*22 inches
       4. 20*30 inches
       5. 22*28 inches
70. Tape – may be celluloid plastic or high tensile polyester film is a ultrathin coating of iron oxide (magnetized)
71. Microphone should be 25 cm away from the speakers
72. In bulletin board – color, harmony and balance is essential
73. In posters to more than 5 words should be used, and arrangement of element should be – logical and easy flowing
74. In diagram, map, chart and graph information is presented in abstracted form
75. Specimen and model – method demonstration, group meeting, training program, exhibition
76. Diorama – exhibition and communication center

77. For specimen at permanent exhibition site display these in glas jars with Bakelite caps
78. Mobile exhibitions – use transparent plastic jars
79. Structural models can be made of - cardboard, Masonite, plastic sheet, light metal
80. Presented type – flash card, slides, filmstrip
81. Display type – poster bulletin, board, model, exhibits
82. in slides devote:
    a. 25 per cent time in introduction the topic and explanation the main point
    b. 50 per cent time in projecting the slides and delivering the talk
    c. 25 per cent time in for question and answer
83. Filmstrip: each filmstrip contain about 30-60 frames (20-36 full frames; 40-72 half frame)
84. Opaque projector; the size of projectable material should be around 25cm * 25 cm
85. Important psychological technique like covering a portion of transparency with a sheet of paper and making progressive disclosure and superimposing diagram may be achieved in overhead projection at the ease and in an effective way.
86. Puppet show may be held at day and night
87. Drama always held at night
88. Films have optical sound tracks which produces sound ad sync with the movement of picture
89. Films to be presented when main extension program is over
90. Video is an electronic audio visual medium for preparing video tapes of program and events which could be projected through a receiving set with moving images and synchronized sound
91. Film and video are the system of recording which result in concrete tangible works

    TV is essential a channel of transmission
92. In motion picture the raw film is coated with photosensitive material which after exposure required dark room processing and printing
93. In video tape is coated with magnetically sensitive material, which is magnetized after exposure and do not require such operations
94. Video tapes have less storage life compared to film tapes

95. Folk media are flexible, thus facilitate and incorporate development oriented message in their themes.

    Help in generating grassroots participation and facilitates dialogue between performer and the audience

96. For developing nations – modern method + folk media
97. I.T. includes both computing and communication technology

    If developed properly IT can provide support to knowledge based extension system

    Users of IT should be careful about the problem of

    a. Urban bias
    b. Information overlord

98. Information system designers must clearly identify

    a. User's goals
    b. The information that users need to achieve the goal
    c. The process by which information will be gathered and entered into the system

    These steps make adoption of IT a 'need drivers' rather than 'technology driven'

99. GIS systematically designed, spatially indexed approached
100. Advantages of IMM (interactive multimedia) are:

    a. It shifts from passive to active learning
    b. Empower learning who can learn at their own pace

101. Two aspects of IMM – 1st creation and 2nd use
102. Farmers and those who serves them must balance formidable objective of

    a. Productivity
    b. Profitability
    c. Sustainability

103. Governance is a concept that encompasses that state institutional arrangement decision making process, implementing capacity and the relationship between government official and public
104. Bhoomi project – Karnataka
105. Gyandoot – MP (community oriented rural internet kiosk), installed in 5 block of district, later 11 more KIOSK were set up

    Entire expenditure of gyandoot is borne by panchayat and community

106. CARD – computer aided registration of deeds – AP
107. FRIENDS (fast, reliable, instant, efficient, network for disbursement of services)
108. E- chaupal net savings – 3 percent of costs
    a. 0.5 per cent goes to sanchalak
    b. 0.5 per cent sanyojak
    c. 2 per cent goes to ITC
109. AGREX: Kerala center for informatics research and advantage
110. Farm advisory system: PAU, Ludhiana
111. Kisan call center – MOA, GOI
112. Cell phones and base station use low power transmitter
113. Digital camera is extremely convenient for use in photojournalism, exhibition, publication and documentation work
114. SATCOM used in both – broadcast mode and interactive mode
115. Expert system in ag. Ext. is essentially a decision support system for farmers and extension agents to obtain expert opinion and advice on specific problems
116. Communication process is governed by 3 level of semiotic rules
    a. Syntactic rules – deals with sign and symbol
    b. Pragmatic rules – deals with sign and symbol and their audience
    c. Semantic rules – deals with sign and symbol and what they represent
117. Process of feedback (for the 1st time) - Schramm
118. Concept of feedback (for the first time) – Berlo
119. According to Berlo – communication is a continuous process in which noise is reduced by feedback
120. Two way model = Westley and McLean models
121. Transactional model – Barnlund (2008) – individual are simultaneously engaging in sending and receiving message or sender and receiver are reciprocally linked to each other. Also known as constitutive model as it focuses in how an individual communication the determining factor on the way message will be interpreted
122. First farm magazine – Kheti
123. National press day – 16th November
124. Block for printing photograph or other illustration or text – cut
125. Interaction – national journal

126. Agriculture and livestock – NIRD
127. Rural India – Pune
128. Journal from IIMC – communicator

Chapter 5

Adoption and Diffusion

1. Adoption is a decision
2. Diffusion is a process
3. Innovation is idea, practice or object that is perceived new by an individual or other unit of adoption.
4. Perception – is an activity through which an individual become aware of objects around oneself and of events taking place
5. Uncertainty is the degree to which a number of alternatives are perceived with respect to the occurrence of an event and the relative probability of these alternatives
6. Form of extension: form is the directly observable physical appearance and the substance of an innovation
7. Function of extension: function is the contribution made by an innovation to the way of life of the people.
8. Meaning of extension: meaning of extension is the subjective and frequently unconscious perception of an innovation by members of social system
9. Attributes: attributes are qualities, characteristic or traits possessed by the object
10. The quality or characteristics of the innovation, is not the intrinsic quality of the innovation but what people thinks about the innovation
11. Perceived attributes of innovation:
    a. Relative advantage

b. Compatibility
c. Complexity – complex technologies often require complementary adoption
d. Trailability – earlier adopters appears to be more concerned about the trialability of an innovation than late adopters
e. Observability
f. Predictability

12. Benefits of preventive innovation are clearly visible after a lapse of timing
13. Preventive ritual is due to lack of knowledge about preventive innovation, or inability to procure them.
14. A preventive innovation has a particular slow rate of adoption because individual have difficulties in perceiving its relative advantage. So, the relative advantage of a preventive innovation is a delayed reward
15. The adoption process:

☆ Ryan and gross – adoption occur in stages (hybrid corn)

Stages:

1. Awareness (A)
2. Conviction (C)
3. Trail (T)
4. Complete adoption (A)

☆ Wilkening (1953) – adoption of an innovation as a process composed of learning, deciding and acting over a period of time

Stages:

1. Awareness (A)
2. Obtaining information (I)
3. Conviction and trial (C and T)
4. Adoption (A)

Wilkening is a decision making process

☆ Johnson and haver – involve following steps (O,A,D,T,A)

1. Observability
2. Making analysis of it
3. Deciding available course of action
4. Taking one course
5. Accepting the consequences of the decision

- ☆ North central rural sociology subcommittee for the study of diffusion of farmers practices. Identified five stages of the adoption.
  1. Awareness (A)
  2. Interest (I): the person what to know what is, how it work and what are potentialities
  3. Evaluation (E):
     a. Whether or not to try it
     b. Mental application of the idea in present and anticipated future situation
     c. Individual judges worth of the innovation
     d. Whether idea applicable to its own situation and if applied what would be the result
  4. Trial (T): test the effectiveness of innovations Considered as practical evaluation of innovation
  5. Adoption (A):
- ☆ Singh says that the stage of adoption are dynamic are not static.

16. Singh's stages
    - ☆ Need (N)
    - ☆ Awareness (A) – person's acquainted with broad features of the society innovation and knows about the source of availability
    - ☆ Interest (I)
    - ☆ Deliberation (D) – mentally examines the idea under his/her own condition. Seek advice of decision maker/opinion leader observe performance at different places
    - ☆ Discuss with family members Take the decision to try or reject the idea
    - ☆ Trial (T)
    - ☆ Evaluation (E) – calculate input-output, risk, uncertainty
    - ☆ Adoption (A)
17. Different kind of communication methods and agents important in each stages of adoption process: (AIETA)
    a. Awareness: mass media > friends and neighbors > govt. source > salesman and dealers
    b. Interest: mass media > friends and neighbors (F and N) > govt. sources > salesman and dealers (S and D)

   c. Evaluation: F and N > govt. sources > salesman and dealer > mass media
   d. Trial: F and N > govt. sources > S and D > mass media
   e. Adoption: F and N > govt. sources > mass media > S and D

18. In NAIDTEA
   a. Need: village level change agent and mass media
   b. Awareness: village level change agent and mass media
   c. Interest: extension agent (formal success) and fellow farmers
   d. Deliberation: informal sources including family members
   e. Trial: no communication, for complex and new processes refer to change agent
   f. Evaluation: fellow farmers and neighbors
   g. Adoption: self-experience at trial stage.
19. High dependence of relatives and friends are negatively related to adoption of new resources
20. Rate of adoption depend is directly proportional to network interconnectedness and rate of adoption for interconnected network take off in a snow ball contagion process

    **** refer to the diagram in G.L. Ray
21. Lippit – outlines condition about why people change, courage to change is the toughest prerequisite to change
22. Distinctive aspect if innovation decision process:
   a. Perceived newness of innovation
   b. Uncertainty associated with newness
23. Stages in the innovation decision process:
   a. Knowledge stages – exposed to innovation existence and gain some understanding.

      Knowledge is initiated by an individual and greatly influenced by one's predisposition.

      NEED DEVELOP MOTIVATION AND MOTIVATION DEVELOP NEED.

      Information seeking in concentrated at the knowledge stage, but it is also occurs at persuasion stage and decision stage.

      When adequate amount of how to knowledge not obtained then rejection and discontinuance is likely to occur.
   b. Persuasion stage: persuasion function is affective mainly related to feeling.

Individual become more psychologically involved and actively seeks information about it

Attributes of innovation are perceived based on one's personality and social system norms; develop a general attitude about innovation

Individual mentally apply the new idea to the present or anticipated future situation before deciding whether not to try it

2 level of attitude:

1. Specific attitude towards innovation
2. General attitude of towards change

At persuasion and decision individual motivated to seek innovation evaluation information At this stage individual form favorable attitude and unfavorable attitude towards innovation.

c. Decision stage:
   a. Individual put the innovation in a small scale trial
   b. Adoption or rejection of innovation occurs at this stage
   c. Demonstration and field days quite effective at this stage

d. Implementation stage:
   a. When individual and other decision making units puts the innovation into use
   b. Where to get the innovation
   c. How to use it
   d. What operational problem are to be faced
   e. How it will be solved

      (Points B to E are the question raised during implementation stage)
   f. Change in management occurs at this stage
   g. Reinvention occurs at this stage
   h. Adoption of innovation is thus a process of social construction
   i. Confirmation:

   When an individual or some other decision making units seeks to reinforce the decision being already made.

24. Rejection happens prior to Adoption and discontinuance occurs after the Adoption.
25. Innovation decision period is actually a gestation period during which a new ferments in the minds of an individual.

26. Rate of awareness knowledge is more rapid than its adoption. Earlier adopters have a shorter innovation decision period than late adopters
27. Rate of adoption is the relative speed with which an innovation is adopted by the members of a social system.
28. Variables determining rate of adoption (rogers)
    a. Perceived attributes of an innovation
    b. Types of innovation decision
    c. Communication channel used or available
    d. Nature of social system
    e. Extension agents promotional efforts
29. Mandates of adoption is the mechanism through which system exerts pressure on the individual to recognize the relative advantage of innovation
30. Over adoption happens due to insufficient knowledge about innovation and inability to predict its consequences. Surveillance, training and communication are the method to prevent over adoption
31. Types of innovation decision:
    a. Optional innovation decision (individual has a complete responsibility for the decision)
    b. Collective innovation decision (individual has a say in decision)
    c. Authoritative innovation decision(individual has no control in decision making)
32. Diffusion effect: the system's self-generated pressure towards adoption also change an increasing proportion of the members of the system to adopt, this increasing pressure from interpersonal network may be termed as the diffusion effect.
33. Persuasion leads to change in overt behavior
34. Three category of consequences:
    a. Desirable vs. undesirable
    b. Direct vs. indirect
    c. Anticipated vs. unanticipated
35. "How to information" is the chief concern in persuasion as well as decision stage. Next important thing is
    a. Yield potentiality or increase in yield – persuasion stage
    b. Availability of inputs – decision stage

36. Adopter categories is based on INNOVATIVENESS

    Innovativeness is the degree to which and individual or other units of adoption is relatively earlier in adopting new ideas than other members of a system.

37. Adoption of innovation follows two type of curves:
    a. Normal/bell shaped – when plotted over time on a frequency basis
    b. S shaped curve (similar to learning curve) – if cumulative number of adopters is plotted

       S-shaped adopter distribution rises slowly at first when there are few adopters in each time period. It then accelerates to a maximum until half of the individuals in the system have adopted. Then the S-curve increases at a gradually slower rate as fewer and fewer remaining individuals adopt the innovation.

38. Adopter categories based on normal curve:
    a. Left of mean minus 2SD → innovators (2.5 per cent)
    b. Between mean minus 1-SD → early adopters (13.5 per cent)

       And mean minus 2-SD
    c. Between mean and mean → early majority (34 per cent)

       Minus 1-SD
    d. Between mean and mean → late majority (34 per cent)

       Plus 1-SD
    e. Right of mean plus 1-SD → laggards (16 per cent)

39. Venturesome, deviate from social norm, deviants, sphere of influence may go beyond community boundaries – innovators
40. Localite, maintain adoption leadership, not to test untried ideas but quick to test tried ideas – early adopters
41. Deliberate, active localite, neighbors and friends from whom majority of the members of the community seek information about innovation – early majority
42. Cautious and skeptical, adopt new ideas just after the average members adopt it, they adopt mainly because people have already adopted and they are getting benefit out of it – late majority
43. Laggards are traditional
44. Extension agent should pursue segmentation strategy, such that benefits of new innovation can be equally distribution among the population.
45. Earlier adopters are not different from late adopters in age

46. From the standpoint of energy expenditure it is better that late adopters should be educated by the fellow farmers rather than to go them by the ext. agent.
47. Management development is suggested by Ray and Bora
48. Bora identified the operational and strategic management attributes of farmers as related to profitability in farming.
    a. Operational management is day to day management
    b. Strategic management is behavioral in nature as it has a bearing on long term management of farm enterprise
49. Operational management attributes:
    a. Abilities to develop production system
    b. Ability to use production inputs
    c. Ability to intensify system operations
50. Strategic management attributes:
    a. Ability to integrate goals of the family and the enterprise
    b. Knowledge of current technology and infrastructure
    c. Self-development through education and work experience
51. Adoption index – Bose
52. Adoption quotient – Chattopadhyay

    Adoption scale is a ratio scale designed to quantify adoption behavior

    Adoption quotient formula is suitable for measuring the level of adoption of farmers pursuing a wide range of technologies.
53. Gap analysis make transfer of technology more specific
54. Critical mass occurs at the point at which enough individuals have adopted the innovation so that innovations further rate of adoption become self-sustaining
55. The benefits from each additional adoption of an interactive innovation increases not only for future adopters, but also for the previous adopters.

    Benefits of interactive innovation flow backwards in time to previous adopters, as well as forward in time to future adopters
56. Peer influence usually makes the diffusion curve take off somewhere between 5 per cent and 20 per cent level of adoption.

Chapter 6

Information Technology

1. M.S. Swaminathan:
    a. Village research center/village knowledge center
    b. Last mile connectivity
    c. Gyan chaupal
2. ARINET – Agricultural and Rural Industries information system NETwork.

   To strenghthen small and micro enterprises
3. LISNET – Land Information System NETwork

   Linking all institution involved in land and water management for agricultural productivity and production
4. ARISNET – Agricultural Research Information System NETwork
5. AGRISNET (2002) – Agricultural, resources information system and networking

   Infrastructure network up to block level agricultural officers facilitating agricultural extension services and agribusiness activity to usher rural property.
6. VISTARNET – agricultural extension information system network
7. AGMARKNET (2002) – network that connect agricultural product from wholesale markets to rural markets
8. VASAT – virtual academy of semi-arid tropics, initiated by ICRISAT, Hyderabad

9. Knowledge management system of NAIP:
   - ☆ Agropedia
   - ☆ RKMP (Rice know management portal)
   - ☆ KVK- net
   - ☆ E- granth: digital access to library resources
     a. E publishing: an open know repository of agricultural resources journal
     b. Krishi prabha: Indian agricultural dissertation repository
   - ☆ CeRA: consortium for e resources in agriculture
10. Knowledge management system of IARI:
   - ☆ Weather based agro advisory services
   - ☆ Pusa agricom
   - ☆ SMS based agro advisory by IARI, KVK
   - ☆ Cyber extension centers in villages State wise cyber extension services:
     a. Karnataka – Bhumi project
     b. Madhya Pradesh – gyandoot (2000), e-Chaupal, Soya Chaupal
     c. Andhra Pradesh – CARD, e-Sagu, e-Seva, FAST
     d. GUJARAT – SEWA, e-Krishi kiran
     e. Kerala – AGREX, pearl, akshaya, kissan, e-Krishi, FRIENDS
     f. UP – harit gyan
     g. TN – RASI, SARI, Agrisnet portal
     h. Rajasthan – e-Mitra
     i. ICAR- READY, ARYA, farmer first

Chapter 7

Rural Sociology and Educational Psychology

1. Sociology – sociology is the study of human relationships
2. August Comte is Known as the father of sociology
3. Sociology originated in USΛ
4. Sociology as a different subject originated in USA
5. Sociology derived from a LATIN word SOCIATUS which means companion or society.
6. Definitions of sociology:

   *** For any definitions it is preferable to remember only the main sentences, or the sentences which are unique

   a. Max weber: science which attempts the interpretations of social man
   b. J.B. Chitamber: study of human relationships in their group relations
   c. Gilin and Gilin: study of interaction arising out of living beings
7. Society – a group who have lived together long enough sharing common values and general interest to be considered as a social unit (J.F. Kuber)

   Definitions of society: (only unique words or sentences)

   a. Adam smith: a society may subsist from different men, without mutual love and affection, if only they refrain to do injury to each other
   b. Ginsberg: connection of individual limited by certain relations and modes of behavior which mark them off others who do not enter into these relations or who differ from them in behavior

8. Both co-operation and disorganization
9. Difference between rural and urban society (Sorokin and Zummerman)
   a. Skill – rural society requires wide range of skills
   b. Urban society requires specialized skills
   c. Size of community – small in rural society
   d. Large in urban society
   e. Social interaction – few and personal in rural society
   f. Numerous and mostly impersonal in urban
   g. Value system - generally sacred in rural society
   h. Generally secular in urban society
10. Man is interacted as a human person in – rural community
11. Man is interacted as name and address in – urban community
12. Less social stratification, but more rigid acceptance of caste differences – rural society
13. Urbanity and mobility are positive correlated.
14. Family – defined by blood relations and sex relationships not by religion or economic group, also known as kinship group.

    Definitions of family: (only unique words or sentences)
    a. Eliot and Merrill: biological and social unit composed of wife and children
    b. Burgess and Locke: group of person united by ties of marriage, blood or adoption
    c. McIver and Page: sex relationship
15. Family is the most common universal group
16. Matriarchy and patriarchy is derived from Greek words, matriarchy is also known as gynoarchy or gynocentric and it is generally found in primitive societies.
17. Book "LOVE AND REALATIONSHIPS" is written by Karol Woityla
18. Types of monogamy:
    a. Social monogamy
    b. Sexual monogamy
    c. Genetic monogamy
    d. Marital monogamy

19. Polyandrous family example – nomadic Tibetans
20. Matrilineal family example – Nair community of Kerala, khasi and jayantia tribes of Meghalaya, bunt community of Karnataka
21. Patrilineal family is also known as – agnatic kinship family, y chromosomal DNA family
22. Group formation promotes – discipline, loyalty, group responsibility and group pressure
23. Primary group are small in size, have responsibility to socialize the individual, primary group are regulated by NORM
24. Secondary group are regulated by formal authority
25. Cliques are found in both formal and informal group
26. Informal group example – play group, friendship group *etc.*
27. In-group-out-group is divided based on – group boundary, which is generally determined by the attitude or feeling of the people towards their own group and towards other
28. Group process is important in group discussion
29. Social group definition – sheriff and sheriff
30. Collectivity: no. of persons whose behavior is specifically polarized towards a temporary center of attraction which leads to unity, as persist as long as the center of attraction exists.

    Types of collectivity:

    a. Crowd: temporary collection of people reacting stimulus.
       - ☆ Casual
       - ☆ Action
       - ☆ Expressive
       - ☆ Conventionalized
    b. Audience: no. of people at the same place, subject to similar stimulus
       - ☆ Recreation seeking
       - ☆ Information seeking
       - ☆ Conversational
    c. Public: scattered group of people who share an interest in particular topic
    d. Social movements: the collectivity getting with some continuity to promote a change or resist a change in a society or group
31. Category: collection of items which has at least one common characteristics

32. Aggregation: temporary collection of individuals in physical proximity ex. Spectators in a cricket match
33. All aggregate are category but all category are not aggregate

    So, aggregate = category + physical proximity
34. Similarly all social group are category, but all category aren't social group

    Social group = category + physiological relations
35. Club: squad is a group of 3-15 people (triad = 3 people)

    Team is an extension of squad
36. Peer group: a peer group in which the members are approximately the same age interest and social status
37. Gang: social unrest
38. Mob: law taken into own hands
39. Primary and secondary group divided on the basis of kind of contact
40. Dwight Sanderson divided group based on structure
41. George Hessen classified based on their relationship with other group
42. Example of horizontal group: religion, nation, political parties
43. Vertical group: small division such as economic class

    Vertical group is a part of horizontal group
44. Small group: less than 25 people
45. Medium group: 25-50 people
46. Large group: 50-100 people
47. Two kind of interest in forming a group:
    a. Self-centered interest
    b. Group centered interest
48. The shared possession of a community sentiment is important
49. Any area of common life like

    Village, town, district, country or even wider area may be regarded as community

    Shared possession of a community sentiment is important

    Community is permanent in nature
50. Definition of community: (only unique words or sentences)
    a. Lundberg: human population living with a limited geographical area
    b. Begardus: we feeling in a limited given area

51. Community is a means whereas it is a source
52. Community are naturally created, not by some act
53. Neighbor is also a community
54. Hunting and agrarian are still considered to be oldest
55. Postindustrial societies are known as – information or knowledge societies
56. Feudal societies is found to be in 16th century
57. Structural and function of society:
    a. Structural elements of a society
        - ✰ Identity
        - ✰ Composition
        - ✰ Intergroup relation
    b. Functional elements of society:
        - ✰ Objective or ends
        - ✰ Norms
        - ✰ Leadership
        - ✰ Resources
58. Socialization is the process by which an individual is conducted into one's social and cultural understanding, socialization involves the development of personality, attitudes, habits and expected social roles.
59. Social system is defined as a set of interrelated units that are engaged in joint problem solving to accomplish a common goal.
60. Determinants of status are:
    - ✰ Wealth
    - ✰ Ancestry
    - ✰ Functional utility
    - ✰ Religion
    - ✰ Biological characteristics
61. Social mobility:
    - ✰ Vertical social mobility
    - ✰ Horizontal social mobility
    - ✰ Geographical social mobility

    An open class system facilitates upward mobility of individuals and groups.

62. Social interaction:
    a. Positive social interaction
        1. Cooperation
        2. Accommodation
        3. Assimilation
        4. Acculturation
    b. Negative social interaction
        1. Competition
        2. Conflict
63. Social interaction is also known as social process
64. Cooperation is a psychological and social phenomenon
65. Accommodation is a type of conflict resolution
66. Form of accommodation which serves as a safety valve for release of hostility – institutionalized
67. A process whereby contending group of individuals strive each other's from achieving the objective, whether or not they desire to secure it for themselves – contravention
68. When member face disagreement and the opposing party wish to maintain a relationship in their own interest – consensus
69. Social institution is formed to meet basic needs of the society
70. Institution are culturally approved pattern of behavior including prescribed roles and procedure, have persisted long enough to be become permanent
71. Family is the most basic social institution, providing procreation, upbringing and socialization Provide economic security, love and affection
72. Religion provide foundations for more of the society
73. Form of govt. and its method of working depend upon accepted pattern and behavior of the society
74. Agriculture is a economic institution
75. Social institution make human being as social animal

    It also provide class status to be individual of the family into which he has been born

    Types of function of social institution:
    a. Manifest
    b. Latent

76. Educational institution: socializing persons into a basic values and practices of a society is a characteristic function of educational institute.
77. Symbolic trait of family: ring, wedding, will
78. Symbolic traits of economic institution: merit award, emblem, trade mark
79. Symbolic traits of political institutions: flag, codes, charter
80. Functions of economic institutions: providing food, clothing, and shelter
81. Social organizations:
    a. On the basis of political structure
       - ✰ Prescribed organization (exist in totalitarian state)
       - ✰ Voluntary organization
    b. On the basis of organization operation:
       - ✰ Open organization
       - ✰ Secret organization
    c. On the basis of admission to membership
       - ✰ Inclusive organization – for everyone ex. Recreational club
       - ✰ Exclusive organization – admission only through selection ex. Rotary interaction
       - ✰ Restricted organization – open to those people who pose certain kind of predetermined qualification. Ex. Association or club of radio, engineers, agronomist, doctors or textile manufactures
82. Counter culture: subculture that adheres to a set of norms and values that are sharply contradict the dominant value and norm of the society.
83. Characteristic of culture:
    - ✰ Culture is learned through – habits, socialization, and thoughts
    - ✰ Culture is universal as well as unique
    - ✰ Culture is static as well as dynamic
    - ✰ Culture has permanence but it changes over time
    - ✰ Culture maintain both continuity and change in society
    - ✰ Ecological diversity is an important source of cultural difference
    - ✰ Culture is idealistic – sum total of ideal norms and values *etc.*
84. Culture change takes place through:
    a. Discovery and invention within in a society
    b. Diffusion and borrowing outside a society
85. Belief, values are the hard point of culture which extension want to change

86. Example of cultural lag – people have increased production by adopting improved technologies but have failed to create organization to ensure profitable processing and marketing
87. Culture is from Latin word cultura, or cicero which means cultivation of soul
88. Culture: continuously changing pattern of learned behavior and products of learned behavior transmitted and shared by member of social system

    It is a typical habit pattern of people
89. In sociology culture denotes - acquired behavior of human beings
90. Culture complex – organization of cultural traits around one dominant trait
91. Three approach of cultural in India – Amartya Sen
    a. Exoticist approach – exoticist approach concentrates on the wonderous aspects and understanding of culture in India
    b. Magisterial approach: sense of superiority and guardian hood. This belongs to various confusion about the culture of India.
    c. Curatorial approach: observe, classify and record diversity of India in different part of India
92. Cultural makes man a human being
93. Culture satisfies aesthetic, moral and religious needs of people of group
94. Cultural diffusion is the process by which a cultural trait invented or discovered in one society and are spread directly or indirectly to other societies
95. Values and belief are culture specific
96. There are values which are considered to be obstacles for change:
    a. Familism
    b. Fatalism
    c. Conservatism
97. Social values are closely associated with NORMS, which have some laws to govern people action.
98. Type of values (franekal)
    a. Moral value
    b. Real value
    c. Market value
    d. Personal value
    e. Instrumental value

99. Social change:

   - ☆ Folkways – informal rules of behavior based on habit and tradition, mostly found in usages, approved forms of behavior for specific situation, this is the expected way in which a person have to be behave

     Folkways arise from experience

     Superstition arises from folkways

     Folkways are the habits and customs of people in a particular culture

   - ☆ Mores: socially acceptable ways of behavior that involve moral standards, group shared understanding about what to do and what not to do in a situation

     Ex. Monogamy, ladies 1st

   - ☆ Norms: established pattern of behavior of social system, it expected that member of a society to follow the norm. A person who deviates the norm is called deviants.

   - ☆ Rituals: it is the pattern of behavior or ceremonies which has become the customary way of dealing with certain situation. Ex. Performing a marriage ceremony

   - ☆ Convention: regulating more significant social behavior, conventions are less violated than folkways. Ex. Wearing clothes in public, eating food with a knife and fork

   - ☆ Customs: socially prescribed forms of behavior which are transmitted by generation. Customs are accepted ways people do things together in personal contacts. Customs are to the groups what habits are to the persons

100. Culture may be thought in terms of three different orders:

   a. Overt culture:

      1. Material culture: products of industry such in implements, vessels and gadgets

      2. Overt behavior: customs, folkways, more

   b. Covert culture: attitude, value, belief *etc.*

101. Three categories of cultural traits:

   a. Universals: which are common to all adult members

   b. Specialties: those parts of culture which are shared only by a certain number of society

   c. Alternatives: traits that are shared by some person but not all of the members of society

Greatest amount of uniformity is in universals followed by specialties

Universals and specialties forms the core of culture, surrounded by alternatives

102. Ethos is related to cultural traits

103. Social participation method of leadership development:

   - ☆ Selection is made by those people who are already holding a position in other organization
   - ☆ Assumed that persons already having active participation in the existing organization will be helpful to provide leadership in the new situation.

104. Leaders hold social power

105. Basic elements of leadership:

   a. Leader

   b. Follower

   c. Social system

   d. Task

106. Functional leader example – agricultural innovator

107. Different type of social power

   a. The elite

   b. Organizational power

   c. Power of unorganized mass

   d. Laws

108. Pattern of leadership:

   a. Group centered leadership

   b. Leader centered leadership

109. Aspects of social change:

   a. Structural change

   b. Functional change

   c. Cultural change

110. Factors of social change:

   a. Biological factor

   b. Physical factor

   c. Technological factor

   d. Cultural factor

111. Factors associated with acceptance of change
    a. Need for change
    b. Desirable for innovation
    c. Proper communication
112. Pattern for change: rogers
    a. Stable equilibrium – no change
    b. Dynamic equilibrium
    c. Disequilibrium
113. Behavior is anything which human does. It has three aspects
    a. Cognition
    b. Affection
    c. Connation
114. Personality – is the unique, integrated and organized system of all behavior of a persons

    Experience + thought + action = personality
115. Types of personality as per C.S. Jung
    a. Introvert
    b. Extrovert
116. Interest are of two types:
    a. Expresses interest- selection and ranking of different activities along with the like dislike dimension
    b. Manifest interest: when a person voluntarily participates in a activity
117. Attitude – Allport Mental state of readiness, pragmatized through experience, exerting a directing and dynamic influencing upon the individual response to all objects and situation which is related. Attitudes are formed in relation to objects, person and values. Attitude are innate and are formed as a result of individuals contact with environments Attitude rooted into motivation and provides a meaning background for individual overt behavior

    Attitude develop from consistency among responses, they are more stable and enduring than opinions.
118. Emotions – state of being moved, stirred up or aroused and involve impulses, feeling, physical and psychological reactions. A negative emotional response may lead to non-cooperation
119. Wish are based on subjective judgment, which prove many time irrational.

120. Prejudice: based on certain assumption generally lead to the formation of prejudice, prejudice is normally difficult to reverse. Suitable enactment of penal provision may reduce prejudice
121. Stereotypes: stereotypes help in knowing how people receivers various group of people or practice or various other social phenomenon Stereotypes may be positive or negative.
122. Rewarding desirable consequences leads to reinforcement of learning
123. Important steps in learning process:
    a. Observation
    b. Analysis
    c. Interpretation
124. Learning through others is known as social learning
125. Principles of learning in farmers field school:
    a. What is meaningful and relevant must be decided by the farmers
    b. Learning is a consequence of experience
    c. Cooperative approaches of enabling
    d. Learning is an evolutionary process (right to make mistakes)
    e. Each person's experience is unique in reality
126. Verbal learning: ability to modify symbols, as in languages makes it possible to learn things
127. Motor or skill learning: involves primarily the use of muscles or body
128. Perceptual learning: as a result of past experience people receive the situation differently
129. Attitude learning: attitude is emotionalized system of ideas, which predisposes an individual to act in a certain way, under certain conditions (helps to develop favorable and unfavorable responses)
130. Solution of problems involves the development of some sort of response to overcome obstacles.
131. Motivation give direction to a activity and selects goals deemed valuable
132. Tasks vary in difficulty from person to person because of
    a. Individual capacity
    b. Experience
133. Rate of progress is a product of [individual capacity and motivation]
134. Reasoning: thinking process is applied to the solution of problem
135. Deviation is developed through socialization

136. Book "motivation and personality" – Abraham Maslow.
137. Components of emotions:
    a. Feelings
    b. Facial and vocal expressions
    c. Cognitive approach
    d. Physical and psychological reason or symptoms
138. Theory of emotions:
    a. Somatic theory of emotion – Williams Jones
    b. Cannon bard theory
    c. Affective events theory – Howard M. Weiss
    d. Two factor theory – singer and Schater

Chapter 8

Program Planning

1. Program – total educational job being done in a particular setting
2. Program – refers to the set of project
3. Extension program – statement of situation, objective, problem and solution. Extension program is relatively permanent but requires constant revision.
4. Program planning: is a decision making process
5. Solution: course of proposed action to change an unsatisfactory condition to that one is more satisfying
6. Project: specification of to be done or procedure to be followed in order to accomplish a particular objective
7. Plan: predetermined course of action. Plan may be tailored to specific projects or they may be established as standing plans (ex. Five year plans)
8. Plan of work: is an outline of activities so arranged as to enable efficient execution of the program.

   It is a statement of activities to be undertaken by individual group of people or organization, within definite stated time to carry out recommendations in the program.

   It indicates what to be done, who is to do it, how it is done, when it is to be done, who are to be served or reached and how the results will be evaluated
9. Calendar of work: plan of activities to be undertaken in a particular time sequences

10. Project management: process of managing and directing the appropriate of resources such as people, materials, money and time to compute a particular project according to the objective set forth

    Or

    Organizing and managing resources to accomplish a particular project.
11. Extension program purpose – improve rural life
12. General objective of having an extension program is to influence people to make changes in their way of life and make a living
13. Objective of extension program: Kelsey and Hearne
14. Concept of need: Leagans
15. Needs are the basis for developing meaningful program
16. Unfelt needs may be brought to the level of felt need through appropriate motivation techniques
17. Ultimate objective of program building is to satisfy the needs of the people
18. Without the plan of work, program remains a theoretical exercise
19. Collection of facts is the starting points of program planning processes
20. Principles of extension program planning – 10
21. Objectives are set based on significant needs
22. Calendar of operations shall be prepared based on the plan of work
23. Seasonal plan/annual plan: for one season or one year
24. "Following through the plan of work and calendar of operation" is one of the most important stages program planning process. Some of the important work at this stage:
    a. Training of participants
    b. Communication of information
    c. Conducting methods demonstration
    d. Making regular visits and monitoring

    Obtaining feedback information as to what is happening to the farmers after the introduction of new technology is extremely important for this stage.
25. Evaluation of progress:

    Three essential steps of evaluation
    a. Setting up of some standard or criteria in relation to objectives
    b. Collection of information
    c. Making judgment and drawing some unbiased and valid conclusion

26. Reconsideration: emphasis is should be on the removal of the technical defects if any, and how to obtain more cooperation and involvement of participants and various organizations
27. Nature of program planning process:
    a. Program planning is an educational process and involve both teaching and learning
    b. Program planning is a unifying and integrating process: through the process of identifying, fixing priorities, establishing objectives and goals and providing continuity and evaluation. People get a total picture of where they are and where they want to reach
    c. Program planning is coordinating process: it reaches coordination of efforts and activities between officials and non-officials
    d. Program planning is an evaluating process: evaluation of accomplishment is not only the criterion. It also provides for evaluation and planning process and the organization.
28. Roles of extension agents are to be performed by the extension agent in a persuasive and consultative manner:
    a. Develop a need for change
    b. Establish an information exchange relationship
    c. Diagnose problems
    d. Create an intent in the clients to change
    e. Translate intent into change
    f. Stabilize adoption and prevent discontinuance
    g. Achieve terminal relationship
29. Functional and financial complementary of disaster and development can ensure sustainable development – Mondal
30. Participatory process if one not careful may reinforce existing pattern of discrimination and disadvantage within rural communities
31. Type of semi structure interview:
    a. Individual interview
    b. Key informant interview
    c. Group interview
    d. Focus group
        ☆ Group interview: used to obtain community level interview, provide access to larger body of knowledge and provide immediately cross check on information

☆ Focus interview: a group of people who possess certain characteristics provide data of qualitative nature of focused discussion

Each interview involves a group of six to eight people who discuss a common topic for one to two hours.

A facilitator is there to moderate

32. Ranking: ranking are useful for sensitive information like income, wealth
33. Preference ranking:
    a. 5 – most important
    b. 4 – important
    c. 3 – undecided
    d. 2 – not so important
    e. 1 – least important
34. Direct matrix ranking:
    a. 4 – best
    b. 3 – good
    c. 2 – not so good
    d. 1 – worst
35. Direct matrix ranking is used to understand the reason for a preference of certain things over the 0ther
36. Wealth ranking is repeated for at least three informants
37. Visualization helps to balance dialogue and it increases the depth and intensity of discussion
38. Map types:
    a. Resource map
    b. Social map: created for residential areas of the village. Wealth ranking household assets survey are done on social map
    c. Health map
    d. Topical map: ex. Aquifer maps done by knowledgeable persons
    e. Impact monitoring maps: village maps, pest incidence, input usage, weed distribution, soil quality and so forth
39. In seasonal calendar daily pattern of activity may be explored by charting typical activity for each hour a day. Seasonal calendar drawn on a linear basis and circular basis.

    ☆ 12 months to show a typical year

    ☆ 18 months to show a change between years.

40. Historical analysis primarily focused on relationships and trends
41. Folklore and songs are valuable resource for exploring history
42. Analysis group discussion:

    Intensive semi structured session in which information gathered in the field is analyzed and recommendations for further action are made.

    Important tool for actively involving community members in discussion.

    Two types:

    a. Innovation assessment: objective here is to develop a priority matrix.

       It helps to assess and prioritize possible options for development activities, according to some criteria which may be

       ✰ Amount of benefit of the community

       ✰ Degree of community participation

       ✰ Sustainability of project

       ✰ Equitability of distribution of benefits *etc.*

    b. Sustainability analysis:

       ✰ Periodic review of the activities without collecting additional information

       ✰ Matrix sharpens the group analytical skill brings out important issues

       ✰ Forced group to ask themselves no. of questions regarding the continuation and modification of program.

43. Participatory evaluation: PE is characterized by maximum community involvement at all level, with emphasis on – cooperation, collaboration and shared resources.

    Participant have shared responsibility, accountability and appreciation of ethnic diversity

    PE is done using a form of self-evaluation and reflection

44. Participatory methods are DISCOVERY LEARNING PROCESSES which do not define problems and solutions for the people, but make things visible so that the people could define their own problems and develop their own solutions.
45. Participation is a learning process
46. Extension work grow out of situation
47. Father of project management – Henry Gantt
48. Scheduling of project management is done by Gantt chart

49. Graphical representation of time required by a persons who is no way associated with its preparation - appraisal
50. CPM is technique of project management - useful in basic management functions (planning, scheduling and control)

    Developed in 1956-57 by Du Pont (USA)

    Activity oriented, deterministic

    Used in construction program
51. Several activity form an event
52. In CPM arrow denotes activity, arrow indicates Energy, time and cost
53. Event is denoted by circle
54. Value of EST and LST in a critical event is same
55. In forward pass the value is considered is - largest convergent EST value
56. In backward pass the value is considered is - smallest convergent LST value
57. PERT: technique used for project involving activities of non repetitive nature.

    Used in defense and research program

    Event oriented, probabilistic and uncertainty
58. Logic framework technique: used for project planning
59. Ideal objective of PRA- empowerment of local people
60. Main innovation of PRA: behavior
61. Insider role in PRA - analyst, presenter
62. Outsider role in PRA - catalyst, initiator
63. Cross checking of data is known as triangulation
64. Main innovation in RRA - method
65. Insider role in RRA - respondent
66. Outsider role in RRA - investigator
67. Ideal objective in RRA - learning from insider by outsiders
68. Techniques of PRA can be divided into three categories
    a. Space: social map, resource map, transect and mobility map
    b. Time: timeline, trend analysis, seasonal diagram
    c. Relation: cause and effect diagram, wellbeing ranking
69. Mobility map: technique used to explore the movement pattern of an individual, a group, or community for various purposes in PRA

Information elicited from the mobility map is: frequency of visit, distance, mode of transport, purpose of visit and accessibility

70. Exploratory PRA: used to understand the broad pattern and trends in the society
71. Modern and tools used for problem analysis
    a. Village introductory meetings: to make an overview of activities that have been have been done during the diagnostic phase and to present priority problems to be analyzed
    b. Problem tree
72. Latest participatory method: PLA (participatory action and learning)
73. PLA – an approach for learning and organizing participation of local communities and groups for interacting with them understanding them and learning from them.

    Changing the behavior and attitude of scientist and development staff (*i.e.* scientist know everything and farmers know nothing) is the principle of PLA
74. Combination of ecological and socioeconomic process leads to – agro ecosystem
75. Approach which uses the key concept of RRA, but emphasizes participation by the village residents and function of external as catalyst and partners for self-determination – PALM (developed by MYRADA)
76. PAR (participatory action and research) – the learning by doing approach in which investigator places his/her knowledge at the disposal of local groups
77. Specific goals makes the objective much more realistic and actionable
78. Unless the strategy is implemented program remain only a statement of wish and hopes
79. A program implies something new
80. 5 level of participation – Biggs and pretty.

Chapter 9

Management and Training

1. Objective of org. determine to a large extent structure and function of organization
2. Goals set down guidelines for organizational activity
3. Goals also constitute a source of legitimacy, since of justifies the existence of any organization
4. Role is a set of activities to be performed by an individual occupying a given position
5. Every formal organization is characterized by various kinds of informal practices, norms and social relationship among its members. These informal practices emerges out over time and represents important parts of any organization
6. Culture as it relates to organization is the general pattern of behavior, shared belief and values members in common
7. Culture is fairly stable and establishes implied rules for how people in organization should behave
8. Approaches of organizational change: (rogers and shoemaker)
    a. Authorative approach
    b. Participative approach
9. Management is the process of designating and maintaining an environment in which individual working together in groups, efficiently accomplish selected aims.

10. Aim of all managers is to create surplus and task of all managers is to secure inputs
11. Management is concerned with productivity; which implies effectiveness and efficiency.
12. SWOT matrix is a conceptual framework and it is done at a point of time
13. Lack of information is due to internal weakness
14. SOWT matrix is done for a particular time. So, strategy designers must prepare several matrices at different points in time for different agro climatic zones and farming system.
15. Four systems of management:
    a. Exploitative authoritative: the leader imposes decisions or subordinates and uses fear to achieve employee motivation
    b. Benevolent authoritative: the leader uses rewards to encourage productivity, but management is responsible and there is no team work
    c. Consultative: the leader listens to subordinates and incorporates some employee ideas, but most subordinate do not feel responsible for organizational work
    d. Participative: the leader engages subordinate, solves problem with team work and everyone feels responsible for achieving organizational goals.
16. Planning is the most important means of communication
17. Planning is basically decision making
18. Social responsibility for extension planning is that their implementation should keep in reducing social disparity and make benefits available to the entire clientele, particularly to resource poor communi9ty
19. At top there is a single or few decision makers who makes an broad plan which may be remain valid for the organization for a longer period of time

    At low level – decision makers may plan for fewer functions, which may remain valid for shorter periods of time
20. At lowest level – they only implement the plans, they don't make any plans
21. Planning results from the interaction of change and client system in context of their – objectives, resources, facilities, constraints, time
22. Contingency planning: Koontz

23. Contingency plan: may be prepared to obtain whatever production is possible in that particular season, giving emphasis on the high level input use to make up the time lost in that season
24. Compensatory plan: generally prepared to increase the production in the next season to overcome the loss occurred in the earlier season
25. Organizing: organizing involves establishing an intentional structural roles for the people to fill in an organization
26. Classical theory of organization developed during: 1900-50
27. Organizational structure:
    a. Should clarify who is to do what and who is responsible for what
    b. Removing obstacles
    c. Furnish decision making and communication network
28. Pattern of distribution of authority:
    a. Line function
    b. Staff function
29. Sources of authority: (LOFT)
    a. Legal: authority granted by law of authority or director of a dept. institution or an organization
    b. Financial: authority to handle finance and financial matter. *e.g.* authority of comptroller, financial officer
    c. Operational: limited to the implementation or operation of a specific task. *e.g.* authority of project director, team leader
    d. Technical: authority which generates from expert or specialist knowledge *e.g.* entomologist, soil scientist.
30. Line function characteristics and examples: independent office and separate identity.

    ex. Directorate of agriculture, directorate of agriculture is the head of dept. and agricultural officers at the zonal, district, subdivision and block level are line personnel.
31. Staff functions: in the directorate of agriculture, ADA, joint directorate of agriculture at the HQ, state level specialist like entomologist, plant pathologist, soil chemist, satisfaction are staff
32. There may be three categories of staff:
    a. Technical: those who deal with technical matters
    b. Officers: those who deal with general administration, personal finance, and audit purchase *etc.*

c. Office staff: who exclusively work in the officers and assist the above and the line personal

33. Span of management is best when the members are 3-8
34. The most advantage of delegation is that the process of delegation stimulates the employees to accept challenge, discover and decide
35. Standing committees are relatively permanent
36. For successful operation of the committee, following point may be looked upon
    a. Purpose
    b. Authority and scope
    c. Size
    d. Members: members have some specialist and few generalist also
    e. Chairpersons
    f. Staff
37. Coordination may be achieved externally by the use of committees and supervisor

    Aim at combining individual, group and organization as 'team' to attain enterprise objective.
38. In three tier PRI, extension agency shall take policy decision with panchayat at the intermediate level

    Focus should be on – employment and income generation in rural areas
39. Input are the material aspects of the technology
40. Common mistakes in organizing:
    - ☆ Failure to clarify relationships. Since, it is the chief source of friction, politics and inefficiency in an organization
    - ☆ Failure to delegate authority
    - ☆ Delegating authority without responsibility which may be lead to organizational anarchy
    - ☆ Multiple subordination
    - ☆ Short circuiting
    - ☆ Giving extra constitutional authority
    - ☆ Formation of clique – narrow exclusive group by a superior with some selected subordinates
41. Staffing: process of selecting, maintaining and developing personnel in position to fulfill organizational objective

42. Training: process of acquiring specific skills to perform the job better
43. In-service training: for the purpose of improving performance of an incumbent holding a position
    - ☆ Promotes professional growth of individual
    - ☆ Problem centered
    - ☆ Learner oriented
    - ☆ Time bound
    - ☆ To broaden their perspective increase capacity to gain knowledge and mastering techniques
44. Induction and orientation training: helps in molding individuals as per the requirement of the job, induction training is related to apprenticeship experience

    Ex. NAARM training for the freshly recruits ARS trainees falls under induction training
45. Foundation training is also known as vestibule training, portal training
46. Maintenance or refresher training: summer institute, training and visit
47. On the job training – periodical or ad hoc training on the job. This training is problem or technology centered
48. Carrier or development training: to upgrade the knowledge, skill and attitude
49. Three phases of training:

    a. Planning phase: at this stage training identification and curriculum development are important

    For identifying training needs: job analysis, task analysis and knowledge and skill gap analysis is important

    Curriculum development: what will be taught – training content?

    a. How it will be taught- training method

    b. Implementation phase: information about training courses made available to the intending person well in time

    c. Efforts should be made to enroll participants to full capacity to the training organization

    d. Evaluation phase: process of determining the relevance, effectiveness and impact of training in relation to its objective
50. Training success depends on designing training system that capture 5 core principle of learning:

    a. Learning is a transformational which takes place over time

b. Learning follows a continuous cycle of action and reflection

c. Learning is most effective when it is address issue relevant to the learner

d. Learner is most effective when people learn in group

e. Learning occurs best in challenging and supporting environment

51. Training is:

a. Learner oriented

b. Jon oriented

c. Need oriented

d. Event oriented

52. Phase in which operational description of the job and understanding organization's receptivity is included – pre-training

53. Training looking into its true senses it is to – stimulate potential

54. Approaches to training extension personnel:

- ✰ Discrete phase approach
- ✰ Process approach (R.D. miller)
- ✰ System approach (York) – based on engineering concept
- ✰ System approach to training – diagnostic approach

55. Training orientation: - six

a. Academic – transmitting contents and increasing conceptual understanding – lecture, seminar, individual reading

b. Laboratory: process of function and change, process of learning – isolation, free exploration and discussion experimentation

c. Activity: practice of specific skills – detailed job analysis and practice with aids

d. Person development: field training, simulation methods, incident and case session and syndicate discussion

e. Organizational development: work with small group to form organization

56. Modalities of training: four

a. Direct contact and distance training

b. Formal and non-formal training

c. Centralized and dispersed together

d. Content and process

Content: knowledge about particular task

Process: how people and things function

57. Phases of training cycle: five
    a. Training need assessment (TNA)
    b. TPD (plan development) – structure
    c. DTP (program designing) - content
    d. TPI (implementation)
    e. TME (evaluation)
58. Training needs are assessed at 6 level of aggregates
    a. Individual
    b. Group
    c. Organization
    d. State
    e. Regional
    f. International
59. Two critical parameters of TNA:
    a. National production priorities
    b. Farmer's felt needs for training
60. Extension personal need assumes three form:
    a. Training need
    b. Personal need
    c. Development need
61. Training must be C,C,R. – comprehensive, consistency, result oriented
62. Dependent primarily on the judgment about the training need methods are classified into:
    a. Rational method: informal talks, observation, complaints, comparison, analysis of report, opinion poll, buzz group technique *etc.*
    b. Empirical method: job analysis, performance evaluation, CIT, FCT, check list or questionnaire
63. Determining training objectives: A,B,C,D (audience, behavior, condition, degree)
64. Training objective should be – S,M,A,R,T (simple, measureable, attainable, realistic and time bound)
65. Training method: arteries and veins

66. Training curriculum: heart
67. Appropriate curriculum can be developed by – concentric circle approach
68. Designing of training program (DTP) have 6 sub-systems
    a. Training objective
    b. Skill mix
    c. Curriculum development
    d. training methodology
    e. supporting training methods
69. extension personnel needs to assume three form: training need, personal need, development need
70. evaluation which develop as an alternative to objective approach: decision oriented to evaluation, also known as CIPP approach

    Context: helps to develop objective

    Input: helps to shape proposal

    Process: helps to guide implementation

    Product: helps to guide recycling decision
71. models of training evaluation – six
    a. Objective oriented training evaluation
    b. Decision oriented training evaluation
    c. Transactional training evaluation
    d. Goal free training evaluation
    e. Adversary training evaluation
    f. Adversary training evaluation
    g. Illumination training evaluation
72. Transactional raining evaluation make the use of ATODJ

    A = antecedents

    T = transaction

    O = outcomes

    D = description

    J = judgments

    Evaluation based on full description and judgments of a program on the ground it has focus on few variables
73. Four level of Kirkpatrick evaluation model: (R,L,B,R)
    a. Reaction – how delegates feel about the training

b. Learning – measure of increase in knowledge
c. Behavior – extent of applied learning back on the job implementation
d. Results – effects of business or environment by the training

74. Experiential learning cycle: (ELC) – James McCaffery

    Four phases: (EPGA)

    a. Experience: actively doing individually or in group
    b. Process: sharing, comparing, processing, reflecting what happened, it is the fulcrum of ELC
    c. Generalization: drawing conclusion, identify general principle, introduction of new theories
    d. Application: planning most effective post training method

75. Seven step model (CLEPGAC):

    a. Climate setting
    b. Learning objective
    c. Experience
    d. Process
    e. Generalization
    f. Application
    g. Closure

76. Spiraling process of experiential learning – David A. Klobb

    ☆ Divergence: high degree of imagination, from Concrete Experience (CE) to Reflective Observation (RO)

    ☆ Assimilation: integration of ideas and formulation of theories is characteristic feature of assimilation, Reflective Observation (RO) to Abstract Conceptualization (AC)

    ☆ Convergence: emphasis of pragmatism rather than conceptualization, Abstract Conceptualization (AC) to Active Experimentation (AE)

    ☆ Accommodation: people like hand on experience, Active Experimentation (AE) to Concrete Experience (CE)

77. According to Kurt Lewin – change in behavior can be achieved through group method

78. Phase of interactive demonstration/phases of learning a new skill.

    a. Cognitive phase – the trainees make attempts to intellectualize the skill which they are to perform

b. Fixation phases – the correct pattern of doing a task are practiced by the trainees until the chance of making errors is reduced to zero. In this phases behavior is conditioned

c. Autonomous phase: the speed of trainees of performing a task is increased

79. Four stages of competence:

    a. Unconscious incompetence: you are unaware of the skill and you lack proficiency

    b. Conscious incompetence: you are aware of the skill but not yet proficient

    c. Conscious competence: you are able to use the skill but only with effort

    d. Unconscious competence: performing the skill become automatic

80. For training organization indicators used is – composite training organization index (CTOI)

81. According to sense, training methods can be developed by – D.G. Reay

    a. Telling

    b. Showing

    c. Doing

82. Normal pace of deliver of words in lecture – 160 words/minute

83. Forum- discussion period which followed after the lecture

84. Brainstorming is originated in USA

85. Convention – representative from local units of parent organizations which have a common interest

86. Conference – assembly of different organization and dept. for discussing issues of common interest end results is a set of recommendations for policy making

87. Seminar is one most important form of GD, which is formal in nature

    Seminar has one or more plenary sessions. This method has advantage of pooling together a large no. of persons, at the end recommendations are arrived at for taking actions.

88. Colloquium: modified version of panel discussion, in which 3-4 resource persons discuss specific topics, in colloquium there is more audience participation.

89. Interaction with audience is not expected in symposium

90. Workshop is a cooperative gathering of individuals who discuss, learn and apply practical skills under expert supervision of an expert

91. Case study: problem situation or a written case is presented to the participants of the program

    To enable them to exercise their analytical power, synthetic and decision making powers

    The interrelatedness of various factors operating in the situation is also highlighted

92. Role playing: successful role playing requires a good empathy; role playing is a particular technique which gives an insight into the complex behavioral pattern of the individual. Role playing units should be small in size so that trainees can really identify with the live roles they play.

93. Brainstorming: spontaneity and creativity is essential, purpose is to promote group creativity. Should be restricted to 20 people per group.

94. Buzz sessions: groups are required to produce ideas, opinions, question in about 5 minutes, purpose is to involve each and every participants, may be used to gather solution to a problem or to gather opinions

95. Sensitivity training: discussion is lightly directed by a behavioral expert who creates the opportunity for participants to express their ideas, belief and attitude.

    Objective of this method:

    a. Increase awareness of their own behavior
    b. How others perceive them
    c. Greater sensitivity to the behavior of others
    d. Increased understanding of group process

    It specifically aims at:

    a. Increased ability to empathize with others
    b. Improved listening skills
    c. Improved openness
    d. Increased tolerance
    e. Improving conflict resolution

96. Phases of team development:

    a. Forming: little agreement, unclear purpose, guidance and direction. Also known as the phase of uncertainty
    b. Storming: conflict, increased clarity of purpose, power struggle, coaching
    c. Norming: agreement and consensus, clear roles and responsibilities, facilitation

d. Performing: clear vision and purpose, focus on goal achievement, delegation
e. Adjourning: task completion, good feeling about achievement, recognition

Training ends here....................

97. Leading: four major ingredients of leading:
    a. The ability to use power effectively in a responsible manner
    b. Ability to comprehend that human beings have different motivations forces at different times and different situations
    c. Ability to inspire
    d. Ability to act in a manner that will develop a climate conducive t0 responding to arousing motivation
98. Leadership is a set of function and a social process
99. Assumed represented type leaders: refers to a persons selected to work with a committee or other leaders because later assumed that he represents another group they desired to work with he may or may not be a leader.
100. Gram sevak is a professional leader
101. Gram sahayak is lay leader/local leader/natural leaders
102. Lassiez fairre leaders are often – rationalizers
103. Rogers designated local leaders as opinion leaders
104. The very idea of training leaders is – leadership can be taught and learned
105. Likert - participative management
106. Work done by others is the most difficult task of motivation
107. A good supervisor should serve as a trainer teacher and a leader
108. Communication is the means by which:
    a. Behavior is modified
    b. Change is affected information is made productive
    c. Goals are achieved
109. Communication is the process by which a proper organization climate is maintained and activities are cultivated
110. Written communication are increased in – secretarial work
111. In time dimension control system can be grouped into three categories:
    a. Control as feedback system – auditing and reporting
    b. Real time information – monitoring and supervision
    c. Feed forward information – budgeting and forecasting

112. Monitoring is done at operational level
113. Monitoring includes collection and analysis of information
114. Monitoring should be continuous process
115. Evaluation provides necessary information for organization health for decision and action
116. Process evaluation: it enables to develop a better understanding of the functioning of the program
117. Outcome evaluation: enables participants to apply the understanding which they develop in the process of evaluation to access the goals are being achieved and how well this being done. It also allows development of performance indicators
118. Steps in evaluation: six
119. Quantitative data are more amenable to statistical analysis, qualitative information facilitates in depth analysis of the situation
120. Supervision is method of Indirect control, more effective than direct control
121. Function of supervision – Appley (1969) – to close the gaps between the desired performance and the actual human performance
122. Function of extension supervision (Benor and Baxter) (1984) – the more important objective of assisting and guiding them to do their assigned task effectively
123. Budgeting provides for pre expenditure control of funds, it is a part of planning process
124. Performance budgeting is recommended by: administrative reforms
125. Purpose of audit: is not fault finding. It provides feedback information to avoid failure and to improve working to the system as a whole
126. Annual report provide some amount of functional control in an organization and also serves as documentation for its work
127. An informal human approach rather than a rigid formal approach may help in better control in the performance of extension personnel
128. Change agent success – rogers
129. Chairpersons of gram panchayat are high in 3 motivations:
    - Need for power
    - Need for achievement
    - Need for affiliation

130. For entrepreneurs:

- ✰ High need for achievement
- ✰ Fairly high need for power
- ✰ Low in affiliation

131. For managers:

- ✰ High on achievement
- ✰ High on power
- ✰ Low on affiliation

132. Gunnar Myrdal chain of capital supply

Increased capital supply → increased productivity → higher real income → higher capital supply

133. Four factors which discriminate regular and defaulter credit users:

   a. Management orientation: the degree to which farmer is oriented towards scientific farm management
   b. Risk orientation
   c. Attitude towards credit institution
   d. Economic motivation

134. Credit under microfinance should be: situation specific, liberal and adequately supplied

135. With credit a process of building of persons self esteem begins that is more important than money itself

136. Right solution lay in helping people to create their own jobs

137. Elements which are the reasons for success of grameen bank:

   a. Group based lending
   b. Taking bank to the poor rather than poor to the bank
   c. Focus on the women
   d. A system of regular weekly installment of the loan repayment.

138. Optimum no. of persons in SHG is 10-20 members.

Chapter 10

Entrepreneurship

1. Entrepreneurship development: process of identifying and utilizing available opportunities and resources to convert an idea into the form of a product or service in the market.
2. Entrepreneurship development represents change in a society
3. Entrepreneur word came from French term entrependre, which means to do something
4. Entrepreneur is an innovator who invariably link up an innovation with the market to earn profit
5. Analysis of a certain situation and taking calculated risks are the starting points of for an entrepreneur to venture into a new enterprise
6. Common characteristics of an entrepreneur:
   a. Intensive drive
   b. Willingness to undertake high risk
   c. Enjoy performing difficult task
   d. Creativity and innovation
   e. Ability to getting things done
   f. Focus on creating value
7. Change as a security born out of self confidence
8. Entrepreneur is an effective:
   a. Communicator
   b. Negotiator

   c. Influencer
   d. Planner
   e. Organizer
9. Kurien combined the principles of:
   a. Principles of cooperation
   b. Extension education
   c. Business management
   d. Dairy technology
10. Mission and vision are the two essential attributes of an entrepreneur
11. Entrepreneurship development: means inculcating entrepreneurial skills required for the setting of operating business units, it is an organized and ongoing process

    Basic purpose: motivate person for entrepreneurial carrier
12. While developing business plan entrepreneur should focus on:
    a. Supplier
    b. Price negotiation
    c. Product cycle
    d. Inventory
    e. Warehousing
    f. Competitors
13. Surveys to prepare developmental plans:
    a. Market survey: market for which goods are intended to be produced – are they local selected districts, state, national or international

       Shall there be a good demand for produce and earn a reasonable profit
    b. Resource survey: are raw material available adequately time and of good quality

       Are suitable infrastructural facilities like water, power, storage, transport, cold chain, cold storage *etc.* can be created without much difficulty, from where resource can be available
    c. Entrepreneurship survey: who would be selected for entrepreneurship development program? Their age, sex, education, family, background, innovativeness *etc.*

       Have leadership management ability

       Have commitment and ability to work and to take risk

14. Opportunity: an attractive and excellent project idea which entrepreneurs searches for and accepts such idea as the basis of his/her investments
15. Steps in identification of business opportunity:
    - ☆ Preliminary evaluation: as soon as the opportunity is decided upon entrepreneur has to evaluate investment opportunities against a set of specific criteria to select those projects which are commercially feasible.
    - ☆ Selection of product or service
    - ☆ Market survey
    - ☆ Control program to collect significant info about proposed venture: offer a package of professional and consultancy services to stimulate entrepreneurial development
    - ☆ Marketing: study the people their needs and preferences before starting any project and design it better than your competitors
16. 1956 – industrial policy regulation of India
17. Factors in favor of small scale industries:
    a. Employment
    b. Equality
    c. Latent resource
    d. Decentralization
18. Small industries extension training: (SIET)
    a. J and K (1972)
    b. AP (1973)
    c. Assam (1974)
    d. Karnataka (1975)
19. TREAD: trade related entrepreneurial assistance and development (it is for women)
20. SIDBI – 1990
21. Stages of growth of small organization – Churchill and Lewis
    a. Existence: organization structure remain informal, entrepreneurs directly supervise the activity, identity of business and the owner is inseparable
    b. Survival: this stage where org. needs capital, who are unable to get capital close the business and make a fresh start
        - ☆ Owner may need one or two trusted people to supervise the activities

- ✰ Here key is to generate cash in shortest possible time to achieve break-even point, to meet the cash requirement for repair and replacement
- ✰ The other need for cash is to fund the company

c. Success stage: company become profitable and has adequate cash flow.

- ✰ Here at this stage entrepreneurs make a decision to leave the company or not
- ✰ A part of functional responsibility can be shared by functional managers
- ✰ Control still remain in the hands of founder/entrepreneurs and the family members
- ✰ Company may decide to remain public or private, but there directors are insiders such as members and close friends

d. Growth stage: risk is associated with this stage.

- ✰ At this stage entrepreneurs decides to use internal financial resources or borrow funds to gain financial strengths for further growth
- ✰ Strategic and operational planning are the key issues which carried out by the entrepreneurs
- ✰ These strategies influence organizational culture.

e. Take of stage: at this stage growth at high rate is focus

- ✰ A formal organizational is established to support various function.
- ✰ Delegation of responsibility to professional managers as a means of systematic approach for vertical and horizontal growth is adopted
- ✰ Organization undergoes transition to become larger organizations
- ✰ Professional also joins the board of directors as enablers

f. Resource maturity stage: that small organization has now grown to out an influence on whole market.

- ✰ Operational and strategic planning and drawing clear line between ownership and management are important at this stage
- ✰ Budgeting and strategy planning

22. Skill development *i.e.* "how to do" is highly important for entrepreneurial training
23. In entrepreneurship development, the knowledge and skill of relevant disciplines including marketing and financial management are

combined together in the form of an integrated action plan to make an entrepreneurship venture successful

24. For quality control of entrepreneur has to reconcile between:
    - Cost of production
    - Customer satisfaction
    - Laws of land

    Quality control should also ensure that enterprise should not generate negative outcome like pollution *etc.*
25. Three maxims of ISO – 9000
    - Document what you do
    - Do what you document
    - Demonstrate what you have done by proof
26. Performance appraisal of sales team are carried out informally
27. Entrepreneurs give high priority to internal communication
28. Entrepreneurs give high priority to self-management
29. Successful business is a commitment to innovation
30. Constant feedback of results as well as setting and adherence of high standard give cutting edge over others
31. Theory of religious belief – weber
32. Theory of entrepreneurial supply – Cochran (emphasize over cultural values, role expectations and social sanctions)
33. Theory of social change – Hagen
34. Theory of group level behavior – young
35. X efficiency – lebienstien - x efficiency is basically degree of inefficiency maintained by a far under imperfect conditions
36. Conceptual model – John J. Kao, this model is also known as ECO model

    E = entrepreneurship

    C = creativity

    O = organization
37. Integrative model: Abdul Aziz Mahmoud
38. Most of the organization is based on division of labor approach
39. Dan Hoff classification of entrepreneurs:
    - Innovative entrepreneurs
    - Adoptive or imitative – treated as an agent of economic development

- ☆ Fabian – skeptical and cautious
- ☆ Drone – orthodox

40. New classification:
    - ☆ Solo entrepreneurs
    - ☆ Active partner
    - ☆ Life timer
    - ☆ Challengers
    - ☆ Buyers
    - ☆ Inventors
41. According to stage of development entrepreneurs can be classified into:
    a. First generation entrepreneurs – who do not possess business background
    b. Classical entrepreneurs – traditional and stereotype in approach
42. According to motivations:
    a. Pure entrepreneurs
    b. Induced entrepreneurs
43. Arthur Cole classification of entrepreneurs:
    a. Rational
    b. Empirical
    c. Cognitive
44. Hagen's creative personality is a mixture of
    a. McClelland high need of motivation
    b. Schumpeter theory of innovation.
45. McClelland needs of entrepreneurship:
    a. Power
    b. Independence
    c. Achievement
46. Desai classification of entrepreneurs:
    a. Business entrepreneurs
    b. Agricultural entrepreneurs
    c. Technical entrepreneurs
    d. 1st generation entrepreneurs
    e. Women entrepreneurs

Chapter 11

Gender Sensitization and Empowerment

- ☆ Gender is a neutral term.
- ☆ **Gender** is a socially constructed roles and responsibilities assigned to men and y women in a given society.
- ☆ **Gender mainstreaming** is the process of ensuring that women and men have equal access and control over resources, development benefits and decision making, at all stages of the development process.
- ☆ Gender Analysis examines women's and men's specific activities, conditions, needs, access and control over resources, access to development and decision making.
- ☆ Sex identifies the biological differences between men and women, such as women can give birth and men provide sperm.
- ☆ Sex roles are universal.
- ☆ ***Gender Bias*** is the tendency to make decisions or take actions based on gender,
- ☆ ***Gender blindness*** is the failure to recognize that gender as an essential determinant of social outcomes impacting on projects and policies.
- ☆ ***Gender awareness*** is an understanding that there are socially determined differences between women and men based on learned behaviour, which affect their ability to access and control resources.

- ***Gender sensitivity*** encompasses the ability to acknowledge and highlight existing gender differences, issues and inequalities and incorporates these into strategies and actions.
- ***Gender Discrimination*** is prejudicial treatment of an individual based on a gender stereotype (often referred to as sexism or sexual discrimination).
- ***Gender equality*** is the result of the absence of discrimination on the basis of a person's sex in opportunities and the allocation of resources or benefits or in access to services
- ***Gender equity*** entails the provision of fairness and justice in the distribution of benefits and responsibilities between women and men.
- Gender Mainstreaming is a process rather than a goal.
- ***Gender equity*** is the condition in which women and men participate as equals, have equal access to resources, and equal opportunities to exercise control.
- ***Gender Relations*** are the ways in which a culture or society defines rights, responsibilities and identities of men and women in relation to one another.
- ***Gender Sensitive*** is being aware of the differences between women's and men's needs, roles, responsibilities and constraints.
- Condition is the immediate, material circumstances in which women and men live.
- ***Position*** is women's place in society in relation to men's. Position involves power, status and control over decisions and resources.
- ***Empowerment*** is the process of generating and building capacities to exercise control over one's life.
- The roles, responsibilities, and activities assigned to women and men based on gender are ***Gender division of labour***.
- ***Gender Analysis*** is the process of analyzing information in order to ensure development benefits and resources effectively and equitably targeted to both women and men, and to successfully anticipate and avoid any negative impacts development may have on women or on gender relations.
- For a gender analysis, all data should be separated by sex in order to allow differential impacts on men and women to be measured, yes
- ***Gender Planning*** refers to the process of planning developmental programmes and projects that are gender sensitive and which takes into account the impact of differing gender roles and gender needs of women and men in the target community or sector.

- ☆ Gender analysis is the first and most critical step forward towards gender-responsive planning and programming.
- ☆ Gender Budgeting looks at the Government budget from a gender perspective to assess how it addresses the needs of women in the areas *viz.*, health, education, employment, *etc.*
- ☆ Coined the term Gender role a john money
- ☆ Women in Agriculture day celebrated in – 4th December.
- ☆ International women's day in – 8th March.
- ☆ Directorate of Research on women in Agriculture is establishing in 1996 at – Bhubaneshwar.
- ☆ First woman deputy Director General of ICAR – Dr. B. Meena Kumari
- ☆ The first global conference on women in Agriculture was organized in March 2012 at New Delhi.
- ☆ The first women Director of Directorate of Research on women in Agriculture – (DRWA) = Dr. Hema Pandey.
- ☆ The first chairperson of Bhartiya Mahila bank – Mr. Usha.
- ☆ Which discipline emphasized the importance of gender – Agricultural Extension.
- ☆ What is the proportion of assets owned by women in the world = 1/100th
- ☆ Thc chairperson of National Commission on Women's – Mrs. Lalitha Kumaramangalam.
- ☆ The first women president in India – Prathiba Patil.
- ☆ The first women chief minister in India – Mrs. Sucheta Kripalani
- ☆ What present of the Agricultural work force is contributed by women in world – 43 per cent
- ☆ Which period attend to women's concerns and priorities in development policy – early 1970's
- ☆ Women SHG, are better consider for building – social capital.
- ☆ The government has set up a committee of experts to prepare the blueprint of India's first women's bank the committee headed by – MBN Rao.
- ☆ International of Women's development Agency (IWDA) established in 1985 at – Melbourne.
- ☆ When the UN women was established – July 2010
- ☆ UN women's – United Nations infinity for gender equality and the empowerment of women.

- ☆ CEDAW? – Convention on the elimination of all forms of Dissemination against women.
- ☆ Which country is the world's female majority parliament with 63.8 per cent parliamentary seats occupied by female – Rwanda.
- ☆ Woman as the companion of man, gifted with equal mental capacity said by – Mahatma Gandhi.
- ☆ The principle agents of food security and household welfare – women.
- ☆ The implementation of the priyadarshini scheme which is a self help group (SHG) based project for holistic empowerment of women and adolescent girls stored in – April 2011.
- ☆ What is the proportion of women in India population – 48 per cent
- ☆ IGMSY.
- ☆ Which is world day of the fight against sexual exploitation – 4th March
- ☆ Rajiv Gandhi scheme for empowerment of Adolescent.
- ☆ Girls – Sable started in – 2010
- ☆ The Government of India was declared as the year of women's empowerment – 2001
- ☆ Protection of women for domestic violence Act 2005 came into effect on – 26th October 2006
- ☆ Which is most accessible financial sources to rural women – micro finance.
- ☆ Protection of women against sexual harassment at workplace, bill passed in – 2010
- ☆ The first Indian women to get magsayray award – kiran bedi.
- ☆ The literacy rate among women and men respectively in India as per census 2011 – 65.5 per cent and 82 per cent.
- ☆ The first women to receive gnanpith Award – Ashapurna Devi.
- ☆ It is striving for equal treatment of women source and to abolishing inequality – feminism.
- ☆ The average labour contribution of women in forming – 55-60 per cent.
- ☆ First Indian women to win the Booker Prize was – Arundhati Roy.
- ☆ Practical Gender needs (PGN) are – short term or immediate.
- ☆ Strategic Gender needs (SGN) are along term
- ☆ ICRW – International Council for Research on women
- ☆ First Indian women president of the Indian National congress was = Annie Besant.

- ☆ Which country declares official National Holiday on Women's Day – Russia.
- ☆ The first chairperson of the national commission for women.
- ☆ When the women youth training extension project was started for Karnataka – 1982
- ☆ Training and Extension for women in Agriculture started – Odisha.
- ☆ Tamil Nadu women in Agriculture was initiated in 1986
- ☆ The first women to receive Bharat Ratna Award – Indira Gandhi.
- ☆ Who was the first women president of the United Nations General Assembly – most Vijayalaxmi Pandit.
- ☆ Self employed women farmers – 48 per cent
- ☆ First global conference on women in Agril. (GCWA) – 2012 – New Delhi
- ☆ National policy for empowerment of women – 2001
- ☆ Womens reservation bill was passed in Rajyasabha – 2010
- ☆ Directorate of Res. on women in Agril – Bhubaneswar, on under ICAR subcentre in Bhopal
- ☆ STEP – programme was aimed to the self reliance and autonomy of women.
- ☆ First international women in Agril. conference @ Australia.

Years and Persons

- ☆ The Employees State Insurance Act, 1948
- ☆ The Plantation Labour Act, 1951
- ☆ The Family Courts Act, 1954
- ☆ The Special Marriage Act, 1954
- ☆ The Hindu Marriage Act, 1955
- ☆ The Hindu Succession Act, 1956 with amendment in 2005
- ☆ Immoral Traffic (Prevention) Act, 1956
- ☆ The Maternity Benefit Act, 1961 (Amended in 1995)
- ☆ Dowry Prohibition Act, 1961
- ☆ The Medical Termination of Pregnancy Act, 1971
- ☆ The Contract Labour (Regulation and Abolition) Act, 1976
- ☆ The Equal Remuneration Act, 1976
- ☆ The Prohibition of Child Marriage Act, 2006
- ☆ The Criminal Law (Amendment) Act, 1983
- ☆ The Factories (Amendment) Act, 1986
- ☆ Indecent Representation of Women (Prohibition) Act, 1986

- Commission of Sati (Prevention) Act, 1987
- The Protection of Women from Domestic Violence Act, 2005
- Janani suraksha Yojana-2005
- Kasturbha Gandhi balika vidyalaya-2004
- Indira Gandhi matritiv sahayoj Yojana-2010
- Rajiv andhi program for empwrmt of adolescent girls- 2011
- Priyadarshini-2011
- Sukanya samridhi Yojana- 2015
- Beti bachao beti padao-2015
- Stand up india- 2016

Millennium Development Goals (MDGs)

- Goal 1: Eradicate extreme poverty and hunger
- Goal 2: Achieve universal primary education
- Goal 3: Promote gender equality and empower women
- Goal 4: Reduce child mortality
- Goal 5: Improve maternal health
- Goal 6: Combat HIV/AIDS, malaria and other diseases
- Goal 7: Ensure environmental sustainability
- Goal 8: Develop a global partnership for development

INDIRA GANDHI MATRITVA SAHYOG YOJANA (IGMSY)

1. A Conditional Maternity Benefit Scheme
2. Conditional Cash Transfers (CCTs)
3. Maternity Benefit in India
4. Incentives to Anganwadi Workers (AWWs) and Helpers
5. Modalities and Implementation

Glossary

Active Adopters: who adopt there an innovation and influence others to do SO.

Active Rejecter: who reject an innovation and influence others to do so.

Active Variable: It is a variable that can be changed or manipulated by a researcher.

Administrator: A person with administrative responsibilities.

Adoption: Adoption is a decision to continue full use of an innovation.

Adoption Categories: Adoption categories are the classifications of individuals within a social system on the basis of innovativeness. The five categories utilised are innovators, early adopters, early majority, late majority and laggards.

Adoption Period: Is the length of time required for an individual to pass through the adoption process from awareness to adoption.

Adoption Process: The mental process through which an individual passes from first hearing about the innovation to final adoption.

Adoption Rate: Rate of adoption is the relative speed with which the members of social system adopt an innovation. Rate of adoption is usually measured by the length of time required for a certain percentage of the members of a social system to adopt an innovation.

Adult Education: Education provided for adults for general educational rather than vocational purposes.

Aesthetics: The Judgement of beauty or taste.

Agricultural Extension: Extending knowledge to Agriculturists. Agricultural Extension is concerned with agricultural education assisting farmers to bring about continuous improvement in their physical, economic and the social well being through their individual and cooperative efforts. It makes available to the farming community the scientific and other factual information, the training and guidance for the solution of problems in agriculture including Animal Husbandry, Gardening, Agricultural Engineering, *etc.*

Aim: This is a broad objective. It is a generalized statement of direction and may have several objectives. It is also said to be an end in view direction to the creative process.

***Analysis of Covariance* (ANCOYA):** Statistical technique used to compare two or more groups on a variable when some other variable may affect the comparison.

***Analysis of Variance* (ANOVA):** Statistical testing technique to determine if variance between the mean on different sets of observations exceeds what may be expected by chance.

Antecedent Variable: A variable, which proceeds the variables under consideration, usually unknown to the researcher at the outset of his study: frequently discovered to be the true causal or independent variable.

Anthropology: A study of man and his works. The discipline has two fundamental aspects: one the study of man as an organism (physical anthropology) and its sub divisions human biology, Primatology anthropromatry and biometrics and two, the study of human societies (ethnology), social anthropology, cultural anthropology, archaeology, enthnography and social psychology.

Aphasia: Inability either to use speech or to understand speech.

Extension job: The job of extension in agriculture and home science is to assist __ farming and home making

Extension service: It means an organization and/or programme for the welfare and development.

Extension work: It means the whole structure of extension work. It includes the process of extension education, the process of teaching and learning.

Innovation system: Innovation can be defined as a new way of doing something ranging from changes in the way we think, to the way we produce new products.

Learning: The development and change in behaviour of a person which occur in relation with the attainment of a certain objectives in called learning.

References

Abraham Maslow (1954). Motivation and personality. Harper and Row. New York.

Bhanot, K.K., *et al.* (1998). Fisheries Technologies for Women. Indian Farming, Oct. 1998: 46-48.

Cairncross, A.K. (1967). Factors in economic development, London.

Chambers Robert (1974). Managing rural development. Ideas and experience from east Africa. Scandinavian Institute of African studies.

Chester I Barnard (1964). The functions of the executive. Harvard University press.

CIFA (1998). National Aquaculture Development Plan, Central Institute of Freshwater Aquaculture, Bhubaneshwar: 60p (mimeo).

Cohen Goel (2004). Technology transfer: strategic management in developing countries. Chapter - 3. Sage publications. New Delhi.

Dhama O.P. and Bhatnagar O.P. 1980. Education and Communication for Development Oxford and IBH Publishing Co. Pvt. Ltd., New Delhi.

Everett m. Rogers (1995). Diffusion of innovation. 4th edition. The free press.

ICAR. 1998. - Information Systems Development: Development strategies and implementation plans under NATP. ICAR 1998.

Jagannath Mohanty (2005). Modern trends in educational technology. Neelkamal publications Pvt. Ltd. Hyderabad.

Katar Singh (1999). Rural development: principles, policies and management. 2nd edition, Sage publications. New Delhi.

Koontz, H and Weinrich, H (1985). Essentials of management. 5th edition, Mcgraw-Hill Book company. New York.

Mehta, S.R. (1984). Rural development policies and programmes. Sage publications. New Delhi.

Merrill, R (1972). The role of technology I cultural evolution, Social biology. Vol. 19. pp: 219-223.

Morris, David (1982). Measuring the conditions of India's poor: The physical quality of life index. Promilla company publishers. New Delhi.

NABARD (1997). National Bank for Agriculture and Rural Development. Annual report.

NDDB (1997). National Dairy Development Board. Annual report.

NIRD (1996). National Institute of Rural Development. Rural development statistics.

Peter F. Ducker (1954). The practice of management. Harper and Brothers. New York. pp: 63.

Ray, G.L. 1991. Extension Communication and Management. Kalyani Publishers.

Reddy A.A. 1971. Extension Education. Shree Lakshmi Press, Bapatla, Adhara Pradesh.

Rodrique, M.V. (1992). Perspective of communication and communicative competence. Concept publishing company. New Delhi.

Rogers E.M. and shoemakers, C (1971): Diffusion of Innovations, The free press, New York.

Shruti, Instant Extension Education New Vishal Publication.

Singh, A.K., Singh, L. and Burman R.R. Dimensions of Agricultural Extension, S.K. Jain, Aman Publishing House, Meerut, Uttar Pradesh.